# ANCHORS

## AND

# ATOMS

**The United States Navy Today**

# ANCHORS

## AND

## ATOMS

The United States Navy Today | by KEN W. SAYERS and NORMAN POLMAR

FOREWORD BY

ADMIRAL E. R. ZUMWALT, JR.,

U. S. NAVY

David McKay Company, Inc.
New York

ACKNOWLEDGMENTS

The authors are in debt to a number of individuals who have assisted in the preparation of this book, especially Mr. Robert Carlisle and Miss Anna Urband of the Office of Information, Navy Department, and Mr. Hoyle A. Taylor of the Bureau of Naval Personnel.

Also, Mesdames Gerri Miller, Rose Sayers, Bonnie Worcester, and Doreen F. Hurley, and Mr. Sand Toler who provided assistance in research, editing, and preparation of the manuscript.

Finally, we are in debt to those Navymen who helped teach us about men, ships, and the sea: Captain D.A. Paolucci, Rear Admiral George H. Miller, Commander Leonard R. Womack, Boatswain's Mate 1st/class Robert J. Egloff, and Gunner's Mate 3rd/class William V. Sayers, all U.S. Navy.

K.W.S.
N.P.

To
Deborah, Mike, Wendy,
and Tomorrow

# Contents

# Foreword

Since the beginning of civilization, man has gone to sea: to fish for food for his family, to engage in the transport of his products, or to wage war against an enemy. Now, in the late twentieth century, the sea is more important than ever, as shown by the emphasis on shipbuilding and fleet modernization in countries which have long been largely land-based.

The U.S. Navy has a proud history. During its almost 200 years of service to the country, our Navy has established a proud tradition. Sea power was a contributing factor to the winning of our independence, and the Navy has been contributing to the cause of freedom ever since.

The Navy has four functions to carry out in the performance of its mission. Simply stated, these four functions are:

- A major contribution to the nation's strategic nuclear policy
- Projection of power overseas
- Sea control
- Overseas presence

Our Polaris/Poseidon submarine-launched missiles are the Navy's contribution to our nation's strategic nuclear policy. Knowing that we have these virtually undetectable weapons on station under water makes potential enemy forces think twice before launching a nuclear attack against the United States or a major attack against our allies.

Second, when directed, we use our projected forces to extend U.S. power into all the world's oceans in support of vital U.S. interests. These forces include our strike carrier forces, with planes capable of projecting power hundreds of miles inland, and our amphibious ships with their ability to land our Marines ashore overseas. Often overlooked, but equally important, is our nation's Merchant Marine, which carries more than 90 percent of the supplies necessary to maintain our Army and Air Force units overseas.

Sea control forces are composed of a complex variety of weapons systems including attack submarines, strike aircraft operating from carriers, maritime patrol aircraft, and surface combatants. These forces deal with the enemy surface fleet and surfaced submarines. Their job is to keep the seas open for us and our allies.

Our final, and perhaps most important mission, is that of overseas presence—the ability to maintain silent forces on the oceans of the world, or to show the flag to reassure our allies and deter potential adversaries.

The forces I have mentioned—the ships, submarines, and aircraft—will all be fully explained to you in the following pages by Norman Polmar and Ken Sayers. Both are highly qualified to do so. Mr. Polmar has studied and written about the U.S. Navy, and he has become one of the foremost authorities on the Soviet Navy. Mr. Sayers is a former Navy officer who blends his personal experience with his writing skills to great advantage in telling stories of the sea.

Please keep one thing in mind as you read this book. With all the complex weapons systems, sophisticated electronic equipment, supersonic aircraft, and deep-diving submarines, it is still our people who make the U.S. Navy great. This was true in the days of "iron men and wooden ships," and it remains true today. Today's Navy men and women are the most highly trained and most capable in the history of the service. Look for their accomplishments as you go through this book.

I am sure you will find the book informative, and I am also certain it will give you a better insight into today's modern Navy and its role in the defense of our great nation.

Good reading.

Washington, D.C. ADMIRAL E. R. ZUMWALT, JR., U.S. Navy
January, 1974                                CHIEF OF NAVAL OPERATIONS

A modern Navy consists of men, ships, and aircraft, as shown here in the aircraft carrier *Enterprise* (foreground), as well as submarines, space satellites, bases, schools, shipyards, and numerous other components.

# Navies 1 in the Space Age

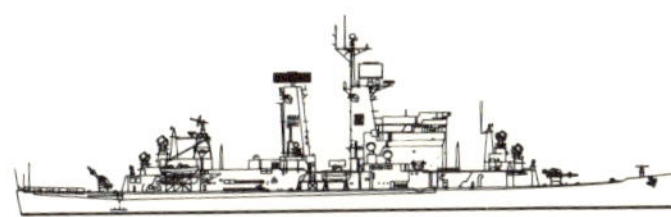

*The Navy of the United States is the right arm of the United States and is emphatically the peacemaker.*

—THEODORE ROOSEVELT (1858–1919)

For more than a decade now we have watched the drama of the Space Age. Since the first successful launch of a man-made satellite, Sputnik I, on October 4, 1957, man has repeatedly shot rockets into the darkness of space.

The United States has now placed scores of astronauts and more than five hundred unmanned satellites in orbit, landed a dozen men on the moon, and fired space probes to the farthest parts of the solar system. This space program has captured our attention and imagination. In contrast, it seems old-fashioned to look at the oceans and the ships and navies which travel them.

Navies consist of ships, airplanes, submarines, shore bases and shipyards, space satellites, and—most importantly—people. All exist to help defend the United States and to support our national interests.

For example, nuclear-powered submarines, armed with long-range missiles, sail beneath the ocean's waves ready to launch a destructive blow against an enemy should he try to strike the United States in a surprise nuclear attack. Aircraft carriers, their escorting ships, and Marine landing units make up powerful conventional military forces for use in political crises or for striking an attacker without the need for nuclear weapons.

Submarines and airplanes perform important scouting or reconnaissance missions—and are capable of seeking out hostile submarines, including those carrying missiles which could hit the United States.

Most of these ships also carry the Stars and Stripes to the shores of allied nations, to show that the United States will stand by our friends or confront our enemies whenever necessary.

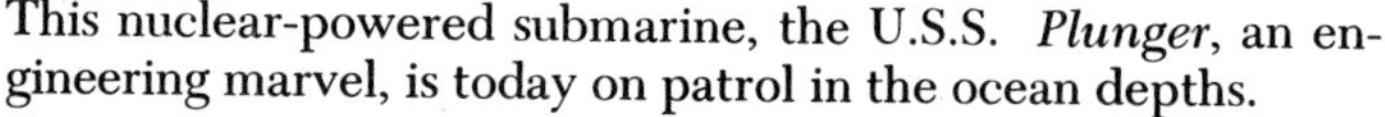

This nuclear-powered submarine, the U.S.S. *Plunger*, an engineering marvel, is today on patrol in the ocean depths.

# THE VALUE OF SHIPS

In all of these activities, ships are valuable because they are:

- — Mobile
- — Independent
- — Difficult to attack

## Mobility

Until you look at a globe or a world map and notice all of the blue area, it's hard to realize that most of the earth is covered by ocean. In fact, almost three-quarters of the globe is made up of water. This is the operating area of the world's navies. Thus ships can range over great areas carrying trade and influencing events over much of the globe's surface.

Take the aircraft carrier as an example. These gigantic ships, longer than three football fields placed end to end, are loaded with fighters, attack planes, helicopters, and electronic and reconnaissance aircraft. When they have to cover a lot of distance in a hurry, the mighty carriers can move more than six hundred miles in one day. They can be in the Atlantic Ocean today and in the Pacific less than a week later.

Because ships can roam around in an ocean, an enemy has a hard time knowing where they will be at any one time. He can't fix their location in advance as he can an air base or a missile launch pad. A navy's mobility gets it where it has to go when it's needed so that "when the going gets tough, the tough get going."

## Independence

Ships sail on the high seas—the free highways of the world. Unlike armies or land-based aircraft, ships are independent of foreign countries because they do not need permission to use international waterways. But ground troops and air units must

The aircraft carrier—seen here from a pilot's view—is a floating base and has become a key component of U.S. naval power. Its large size and high cost, however, have made it the target of considerable criticism.

seek permission from another nation whenever they want to use that country's territory for bases or travel.

Therefore, it is easy to send a naval task force into an area to "show the flag" or to help a friendly nation. Such a force need not obtain an advance OK from nearby countries, and it can be "visible" as much or as little as desired. There is little risk to U.S. military personnel afloat and to civilians ashore.

## Survival

Ships are hard to kill. While at sea, they are on the move and are beyond the range of most enemy aircraft. They cannot be "sighted-in" by land-based, long-range missiles. Ships have side "belts" made of steel, armored decks, airtight rooms called compartments, and equipment to limit and repair damage.

Naval history is filled with cases of carriers taking fantastic punishment from bombs, torpedoes, gunfire, and even suicide planes while remaining afloat. Small destroyers have been struck by crashing aircraft or have had their bows blown off by torpedoes, and they have still fought on. A ship's crew seals off a damaged compartment to keep fire or flooding confined in just

"A destroyer is a lovely ship" wrote author John Steinbeck. These destroyers seem more businesslike as they maneuver "close aboard," personifying the Navy's readiness.

one section. An airplane, missile, or tank—or a foot soldier—does not have that kind of protection!

# USE OF THE SEA

A strong navy—made up of ships with great mobility, independence, and an ability to survive—prevents an aggressive nation from interfering with the American use of the sea. This is most important because the use of the sea is vital to the survival and growth of our country.

Today almost all of the world's international commerce—99 percent to be exact—moves in ships. Even though we are living in a time when jet planes carry passengers between continents, cargo still goes the same way it has for centuries—by ship. Why? Airplanes can't haul the millions of tons of oil, gas, and other materials which are needed to feed U.S. industry. Ships also carry finished products from American factories, plants, and farms,

These three warships, the guided missile frigate *Bainbridge* (foreground), the missile cruiser *Long Beach* (center), and the aircraft carrier *Enterprise* are nuclear propelled. They have steamed around the world in sixty-four days, traveling 30,000 miles, and using only a few ounces of uranium fuel.

The sail frigate *Constellation*, built in 1797, survives as a relic
and technically remains in commission.

overseas to foreign buyers. The waterways are two-way "streets"
for U.S. commerce.

Ships also carry a large amount of our strategic materials.
These are the resources which are required by our nation to
provide for American industrial, military, and naval needs.
Currently, there are seventy-one items on the government's list
of strategic materials—and sixty-nine of them are imported by
ship in various quantities. These include iron ore, natural rubber,
nickel, zinc, and lead.

The life beneath the waves—fish and marines plants—are
becoming more important to mankind. Fish are rich in animal
protein, and the demand for protein is increasing along with the
earth's population.

Thus, the United States depends upon the sea more and
more as time goes by. Every baby or light bulb added to the

The U.S. Navy has been preeminent on the seas for twenty-five years since World War II. Now it is being challenged by the Soviet Navy. Here U.S. and Soviet naval officers chat during a rare meeting aboard a Soviet missile-armed destroyer in Massawa, Ethiopia.

nation requires additional forms of energy; the baby wants food and the light bulb needs power. The food and the power are produced from materials which are either found in or are carried over the sea.

Throughout history and right up to today, the U.S. Navy has safeguarded the American use of the sea. Because of that, U.S. cargo ships, tankers, and fishing craft have helped the country become prosperous and strong.

The Navy is at one time both traditional and modern. Warships sailing in blue water around the world follow time-honored customs, but they are packed with the latest electronic equipment and missiles. Be they powered by steam, diesels, gas turbines, or nuclear energy, today's naval ships are no more out of date than are the ocean's themselves. This is an era in which men of all nations are very dependent upon the seas for commerce, political influence, energy supplies, and defense. This dependence translates into a need for navies.

Navies are of great value to nations. This small volume is an effort to describe the U.S. Navy—past, present, and future. This book was written as the U.S. Navy experienced probably the most dramatic changes since the end of World War II.

The end of the draft and adoption of the All Volunteer Force for providing men and women for the armed services had a considerable impact on the Navy in the early 1970s. Previously, the draft led many to enter the Navy for a more adventurous and educational career.

Also, several technological developments will change the Navy during the next few years. These include fixed-wing aircraft that take off and land straight up and down; ships that fly over the sea on small submerged wings called hydrofoils, and other ships that ride over the water on a bubble of air; improved missile-armed submarines; satellites for navigation and communications; and lasers that can focus intensive beams of light for eye operations and for destroying aircraft.

These changes are producing a modern Navy—a fleet for the 1970s. Thus, this book addresses the Navy's men and women, ships, submarines, and aircraft, and their significance to the continued economic growth and security of this nation.

The *Alfred* was typical of the ships that made up the rather haphazard Continental Navy. The first commander of this ship was John Paul Jones and, carrying thirty cannon plus a few small guns and mortar, she was one of the most-powerful Colonial ships.

# The 2 United States Navy

*Naval force can never*
*undermine our liberties.*

—THOMAS JEFFERSON (1743–1826)

Even before the formation of the United States, the Navy and its ships have been the protectors of American liberty. Indeed, beginning with the first trips to colonize the New World in the seventeenth century, ships were essential to the thirteen colonies. They brought settlers to these shores, served as a link to the mother country, provided for overseas trade, and became the basis for one of the first American industries—fishing. In a sense, therefore, the colonies were a product of the sea.

## REVOLUTIONARY WAR

Despite this heritage, the patriots had *no* warships to help them when they began their fight for independence from Great Britain. On the other hand, the British had some 270 combat ships in their fleet, including 131 ships of the line, the battleship of the day.

General George Washington, the 43-year-old commander of the Continental Army, soon realized he needed a navy. In July 1775, his troops were laying siege to British-held Boston. Washington could not attack the British because his forces were short of cannon and gunpowder. But the British didn't have to sit in the trap without supplies because they brought in men and material by sea. So the American general came up with a plan to plug the hole in the siege and to capture British supplies for his own use. Washington decided to take to the water by using ships with ground troops as crews. The schooner *Hannah*, a small sailing craft, was the first member of "Washington's Navy." On September 7, 1775, while on her maiden voyage, she captured the British ship *Unity* as her first victim.

This experience led Washington to ask the new Continental Congress for naval ships. The Congress heard the call and established a Naval Committee on October 13, 1775, appointed Esek Hopkins as the fleet's commander less than a month later, and added thirteen ships for the "Navy of the United Colonies" on December 13.

But before the new navy could make a solid impact on the war, Washington once again saw sea power at work against him. In March 1776 he still had most of the British forces sitting tight in Boston. His troops brought up cannon to the heights overlooking the city from the south. The British did not want to be trapped under the higher guns and so they slipped from Washington's grasp—by sea. This would have been impossible if the Colonial forces had been able to mount a sea blockade of Boston Harbor. Unfortunately, the Redcoat army that got away came back to sting the Americans after a landing at New York —again, by sea.

Very early in the Revolutionary War, Washington and the Colonials had seen the value of sea power. They realized that there was a large advantage to being able to move forces along the coast by ship, either to attack the enemy where he was weak or to withdraw from him when he was stronger. They also saw the need to interfere with Britain's merchant marine because the island nation was dependent upon the sea for its commerce and for the maintenance of its growing empire.

On March 23, 1776, the Continental Congress responded to the need by authorizing privateers. These were privately owned ships, sailed by civilian crews, which were given license to attack an enemy's merchant vessels. Over the course of the war some 500 American privateers sank or captured nearly five times that number; in fact the British lost 625 ships in just one year (1781) alone.

The sea war also produced the earliest traditions of the United States Navy. One of the most famous Navymen in history, John Paul Jones, brought America's fight home to the British by his brilliant raids against British shipping off England and his daring landing ashore on British soil. Jones showed that youth is no handicap to ability by commanding a merchant ship at age nineteen. Later, in September 1779, he was skipper of the Continental Navy frigate *Bonhomme Richard.* In an epic night battle with H.M.S. *Serapis,* off Flamborough Head, Jones's ship was shot full of holes and was blazing in the dark. Captain Richard Pearson of *Serapis,* thinking the American had hauled down his colors, yelled across the water to Jones to see if he had surrendered. Jones roared back: "I have not yet begun to fight," a proud reply which has inspired generations of Navymen since. He was true to his word, because he brought the *Bonhomme Richard* alongside the British ship, and his valiant crew fought hand to hand and defeated their Royal Navy opponents.

The British suffered a larger and final defeat in the autumn of 1781. France had joined the fight on the side of the Colonies three years before, and had sent land troops and a naval force to America. By October 1781 a combined ground army led by Washington and a naval squadron under French Admiral de Grasse squeezed British General Cornwallis between land and sea at Yorktown, Virginia. Realizing that he was trapped with no place to run, Cornwallis surrendered his 7000 troops to the

America's first naval hero was John Paul Jones. Born in Scotland and going to sea at the age of thirteen, he became the outstanding sea captain of the American Revolution. In 1779 he commanded the *Bonhomme Richard*, a leaky, rebuilt French merchant ship, in the epic sea fight with the British warship *Serapis*. In the heat of battle the British captain asked if Jones surrendered; his reply: "I have not yet begun to fight!"

Continental Army. This brilliant and well-timed stroke collapsed the British campaign. A treaty of peace was signed in Paris two years later.

Soon after the Revolutionary War, the Congress decided not to support a navy, and by 1785 all U.S. warships had been sold. But before the end of the eighteenth century, Congress had to reconsider this decision because French Navy ships were interfering with American ocean commerce, and Barbary pirates were capturing U.S. merchant ships along Africa's northern coast.

## SMALL WARS WITH FRANCE AND PIRATES

In 1798 the United States had a mini-war with France, our former ally at Yorktown. The French had been angered by the U.S. refusal to protect their New World territories from Britain (this had been agreed to in a 1778 treaty). So, Paris ordered the seizure of American ships and cargoes on the high seas.

But unlike the bleak days of 1776, the United States did have a few naval ships on hand—but just a few. The Navy entered the war with only three frigates: the *Constitution* (with forty-four guns), the *Constellation* (thirty-six), and the *United States* (forty-four). These ships, and others later ordered by Congress, captured 111 French privateers and sunk four more. France finally made peace in a September 1800 treaty in which she accepted American neutrality and agreed to stop interfering with U.S. vessels.

Meanwhile the Barbary pirates of North Africa were capturing merchant ships and enslaving their crews. After first trying ransom payments and diplomacy, without success, the United States declared war on Tripoli, one of the pirate states, in February 1802. Several skirmishes in the Mediterranean, including the famos raid by Lieutenant Stephen Decatur to destroy the captured American frigate *Philadelphia*, finally overwhelmed the pirates. The Yankee merchantmen were finally able to sail the warm Mediterranean without fear.

## WAR OF 1812

After battling with France and then with the ferocious Tripolitan pirates, the United States Navy next had to fight England again. Britain was waging war against Napoleon, and, short of manpower for its warships, it began forcing American merchant seamen into the Royal Navy. Congress declared war against Britain in June 1812.

Very early in the conflict the U.S.S. *Constitution,* skippered by Captain Isaac Hull, earned the famous nickname of "Old Ironsides" in a sharp and successful fight with the British frigate *Guerrière.* Later, the British tried a replay of their Revolutionary War tactic; they blockaded the East Coast and put troops ashore. In August 1813 British soldiers sailed up the Patuxent River near Washington, landed, and marched on the city. They managed to set fire to the Capitol, White House, and Treasury Building. This plainly showed the destruction that a sea-based force could cause if it operated in coastal areas and projected its power ashore. Up on Lake Erie the following month, U.S. Commodore Oliver H. Perry battled the British with a flotilla of homemade boats. A small squadron under Navy Commodore Thomas Macdonough met a southbound British push from Canada on Lake Champlain. Although he was outgunned ninety-two to eighty-six, Macdonough stopped the thrust during a bloody fight off Plattsburg, and the British turned around and marched back to Canada. Another small U.S. Navy squadron at New Orleans stalled a British landing in 1815, allowing U.S. General Andrew Jackson to gather his forces and score the final victory of the war. Ironically, the British had agreed just fifteen days earlier to stop the war.

## CIVIL WAR

The tragic War Between the States was as much a sea conflict as it was a land war. Because of its strength at sea when the Civil War began in April 1861, the North decided to cut the South off from its commercial trade with Europe by a naval blockade. The plan was to strangle the Confederacy in a death lock by using the Union Navy to control the seas and inland waterways, split the South along the Mississippi River, and to support land operations with gunfire and assaults along the coasts and riverbanks.

The Union blockade was effective. The North built up its

The frigate *Constitution* fought against both French and British ships to earn honors and the nickname "Old Ironsides." American-built and manned, she demonstrated the new nation's prowess at sea. The *Constitution* survives as a relic at the Boston Naval Shipyard.

navy (it had eight, and later nine, of the nation's ten navy yards). It stopped the South from shipping out cotton and from bringing in war supplies. It attacked Rebel forts up and down Southern rivers and harbors. Despite heroic efforts by the Confederate

The first combat between iron warships was the encounter between the Union *Monitor* and the Confederate *Virginia* (née *Merrimack*) at Hampton Roads, Virginia, in March 1862. The fight demonstrated that wooden ships could not survive against ironclads and began a new era in naval history.

armies—General Lee's troops reached all the way into Pennsylvania—the Southern effort was doomed by the lack of military material. The Union's sea power drained the South of strength and outflanked the Confederacy by giving the North access to its interior lines. Peace came to the shattered country in April 1865.

The March 1862 fight between the North's *Monitor* and the South's *Virginia* (formerly the U.S.S. *Merrimack*), was history's first contest between ironclad ships. When the Union Navy completed its wartime job, it had become one of the world's leading naval forces in terms of technical improvements and the number of ironclads. But shortly afterward, the Navy went into one of its regular postwar declines, and the European powers pushed ahead with their own ship designs.

## SPANISH-AMERICAN WAR

The United States was poorly regarded as an international naval power in 1898. In that year the U.S. Navy got the chance to prove it had matured into a first-string fighting force because

Dents in her iron turret mark damage to the *Monitor* after her historic battle with the *Virginia*. The Confederate ship was not sunk in the fight, but clearly was the loser and subsequently was sunk by her crew. The straw hat on the officer at right was a standard part of the official uniform in 1862.

The battleship *Maine* steams into Havana harbor. Her sinking in that port in 1898—from unknown causes—helped spark the Spanish-American War. That conflict was largely a naval conflict, with Dewey, Sampson, and Schley commanding the sea forces that freed the Philippines and Cuba from Spanish domination.

The 1800s saw numerous technological advances in ships, among them iron hulls and armor, steam propulsion, screw propellers, rifled cannon, and torpedoes. Here the small, 31-ton torpedo boat *Stiletto* fires a torpedo during evaluation of the "underwater" weapon in the 1890s. Torpedoes are still a primary weapon of ships, aircraft, and submarines.

it had to duel with the Spanish Navy after Congress declared war against Spain. Commodore George Dewey quickly set out to show what our Navy could do. He demolished the Spanish fleet in the Philippine Islands on May 1, 1898. The Spanish lost every ship at Manila Bay and one-quarter of their men were casualties. In contrast, not one of Dewey's ships was damaged! In Atlantic waters the American fleet blockaded Santiago, Cuba, pinning down the Spanish squadron of four armored cruisers and six torpedo-boat destroyers. On July 4, 1898, the Spaniards steamed out to meet the waiting American gunners and they were quickly mauled during a running sea battle. By the following year the

U.S. Navy had emerged as a major sea power on two oceans. About this time, too, the United States, along with Britain, was developing the dreadnought, a battleship armed with large-caliber guns. Powered by steam and fueled by coal, the early battleships and battle cruisers—symbolized by President Theodore Roosevelt's Great White Fleet of 1907–1909—confirmed the new American strength at sea.

In little more than 130 years, the United States had grown from thirteen colonies tied to Britain into a sprawling, industrial society of forty-six states stretching from the Atlantic across a continent to the Pacific. The nation's growth during that period was due in part to the Navy's ability to guard our vital ocean commerce and attack our enemies whenever they threatened U.S. freedom. But the confrontations of America's first century were to be surpassed by the challenges of the second one hundred years. During the twentieth century the U.S. Navy had to fight in two world wars and participate in a long and confusing Cold War. The next two chapters will look at these developments.

Two examples of the German U-boat of the World War I era. These small, uncomfortable craft, which could dive underwater for a few hours at a time, almost defeated Great Britain.
*Imperial German Navy*

# THE 3 WORLD WARS

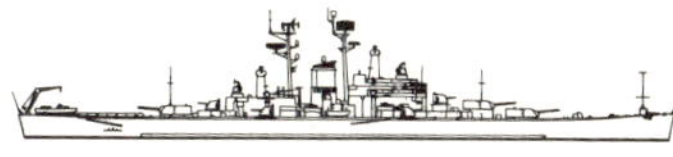

*We make war that we may live in peace.*

—ARISTOTLE (384–322 B.C.)

The United States Navy grew to maturity during the twentieth century's two world wars. Called upon to defend, first, the freedom of other countries, and then U.S. liberties, the Navy met the test all around the world. Twice America assumed the burden of global leadership and international warfare and twice the Navy carried much of the load.

As the 1900s began, the Navy was being modernized. Large big-gun, steel-hull ships were going to sea, and the first U.S. submarines and destroyers—two new important ship types—were joining the fleet.

## WORLD WAR I

When war swept across Europe in 1914, the stormy Atlantic kept the fighting away from America's shores. The United States, separated by 3000 miles of ocean, took a neutral position although we were sympathetic to the British and French. Britain, meantime, faced crippling problems because of a low food supply. This was brought about because she had lost the use of the sea to the German submarine (called *Unterseeboot* or U-boat).

The 140 U-boats operating in the Atlantic were mauling British merchant ships. Late in 1916, for example, U-boats were destroying about 37 ships a month. In February 1917 they sank 86 vessels, while in March the number lost shot up to 103. This no-holds-barred submarine war finally forced President Woodrow Wilson to act. All the free nations of the earth, he said, had to stop Germany "to make the world safe for democracy." Thus, the United States declared war against the Kaiser in April 1917.

Our European allies needed U.S. supplies and troops to defeat the Germans, and the men and material would have to cross the U-boat-infested Atlantic by ship. With many merchant ships already lost to German torpedoes and guns, America's Navy would have a big job on its hands.

A convoy of Allied merchant ships and troop transports crosses the Atlantic with destroyers "screening" the convoy against German submarines.

U.S. Navymen on the open bridge of a destroyer look out for German U-boats while escorting a convoy to Britain. During the war, sound listening devices called ASDIC, and hydrophones, and later sonar, were introduced for finding submarines.

The Navy escorted the merchant ships across the ocean in groups called convoys. Destroyers and subchasers steamed along the front and sides of the convoy, like blockers and flankers, to attack prowling submarines. The British wanted both American escorts and battleships, and the Navy sent both to our allies. United States battleships never fought in combat against the enemy (in fact, there was only one fight involving British and German battleships), but the destroyers were well used during the war. By the end of the conflict in November 1918, some 300 U.S. warships manned by 75,000 officers and men were in European waters.

The U.S. Navy saved many lives during the war. More than two million American troops were transported across the Atlantic to France, scene of most of the land fighting, without the loss of a single doughboy to submarines. Without these troops and U.S. supplies, the Allies would have been locked in a stalemated bloodbath for many years. Instead, American soldiers, sailors, and material turned the tide, and Germany was defeated just a year and a half after the United States entered the war.

## BETWEEN THE WARS

The U.S. Navy built up during World War I until it was the largest in the world. But after the treaty of peace was signed, the United States, Great Britain, France, Italy, and Japan decided to reduce some of their armaments. In 1922 we agreed to scrap fifteen old battleships and eleven other large warships then under construction. This trend was reversed a few years later; in 1933 the Navy was allowed to start building two aircraft carriers, four cruisers, and twenty-four destroyers and submarines.

The aircraft carrier, which began appearing in the 1920s and 1930s, was a revolutionary step forward. Both the British and Americans were discovering the advantages of basing aircraft at sea. By loading bombs and torpedoes aboard the carrier's planes, a naval commander could strike at the enemy's fleet beyond the range of his own guns. He could send out airplanes to scout for the enemy and learn of his location and strength. But at the dawn of World War II, battleships were still the Navy's main weapon. It would take four years of warfare in two oceans before the battleship was retired as queen of the seas.

## WORLD WAR II

Just two decades after the "War to End All Wars," men again took up arms against each other on a global scale. The Japanese moved against Manchuria in 1931, Italy attacked Ethiopia in 1934, and Germany started a *blitzkrieg* war in Europe in September 1939. In the beginning, the Pacific and Atlantic oceans kept the United States out of the growing conflict. But with German bombs pounding English cities and factories, and with Hitler's U-boats torpedoeing British ships, President Franklin D. Roosevelt moved to head off a Nazi victory.

Fearful that the U.S. Pacific Fleet would interfere with their aggression in the Far East, Japanese carrier planes struck Pearl Harbor on December 7, 1941. This view shows Ford Island in the center of the harbor just as the attack began with Japanese planes attacking U.S. battleships moored on the far side of the island. *Imperial Japanese Navy*

He sent U.S. Navy destroyers on "neutrality patrols" into the mid-Atlantic to discourage the U-boats. Under the Lend-Lease program, Roosevelt gave the Royal Navy fifty old American destroyers in exchange for base rights. But soon America was brought directly into the war by Japan.

Using six carriers and about 350 planes, Japan struck at the giant U.S. base at Pearl Harbor, Hawaii, on December 7, 1941. In less than two hours the mighty Pacific Fleet lost four of its battleships, 188 aircraft, four other warships, and 2300 men. Luckily, there were no U.S. carriers present in Pearl Harbor on that Sunday, and Tokyo would soon regret the failure to destroy

United States battleships burning and sinking at Pearl Harbor on the morning of December 7, 1941. Of eight U.S. battleships in the harbor, four were sunk, another run aground to prevent sinking, and the three others damaged. Total Japanese losses were twenty-nine carrier planes and a few two-man midget submarines.

the flattops. Still, Japan had dealt the United States one of its worst military defeats in history, although the United States ultimately won the war four years later.

Many historians regard World War II as an air war and in many ways it was. British and American bombers smashed Germany into ruins, and U.S. Army Air Force B-29 Super Fortresses burned out Japanese cities. But how did those bombers get their fuel and bombs in the first place? How did the troops who fought in Europe and the Pacific get there? How were they supplied?

The simple fact is that the air campaigns of the Second

The destroyer *Shaw* blows up during the Japanese attack on
Pearl Harbor.

World War resulted from the use of the sea by the United States
and our Allies. Every bomb, bullet, and gallon of fuel used by the
U.S. Eighth Air Force based in Britain was carried across the
Atlantic by merchant ship. In a sense the bombs which fell on
Germany began their "flight" in the cargo holds of a ship, were
carried 3000 miles over the ocean, off-loaded and shipped by rail
or truck to an air base, stacked in the belly of a bomber, and
flown only the last 400 miles by plane.

Also, the fighter aircraft needed to escort the bombers to
their targets were carried in ships. The U.S. Navy, along with the
Coast Guard which became part of the Navy during the war, and
the British and Canadians, stopped the U-boat "wolf packs" cold.
Thanks to the destroyers, destroyer escorts, subchasers, and escort

carriers, the German hunters soon became the hunted. They were chased from the Atlantic, making that ocean safe for merchant ships.

The Western Allies soon went on the offensive. The first landings in North Africa in 1942 were followed by the invasions of Italy in 1943 and of Normandy, France, in 1944. These assaults required thousands of tons of tanks, trucks, gasoline, ammunition, food, and troops which had to be transported from the United States out to the advanced bases and even right to the landing zones—in ships. The Soviets, who were fighting Hitler on Germany's eastern flank, received most of their supplies from the United States.

Across the world in the Pacific, the Allies—mainly the United States—used surface ships and submarines to destroy the Japanese Navy and merchant marine. Japan, like Britain, was

This was part of the invasion of the Philippines by American forces in 1944. World War II saw amphibious landings on a scale beyond imagination. Names such as Normandy, Iwo Jima, Guadalcanal, and Okinawa, where sea power permitted direct assault against the enemy, are now names basic to American history.

Among the less glamorous but most vital ships operated by the Navy and Coast Guard during the war were the LSTs, tank landing ships, which could land tanks, trucks, guns, and troops on the enemy's beach. These Coast Guard-operated LSTs are landing Army troops in the Philippines.  *U.S. Coast Guard*

highly dependent upon ocean traffic, and when this dried up, Japan was in trouble. U.S. amphibious forces began sweeping westward over the ocean in 1942. Soon, the United States was able to build airfields on some of the captured islands, and the big B-29s started winging their way to Japan. Just as in the European war, the materials for building these airfields and arming the bombers had to be ferried across the sea by ship.

A U.S. destroyer approaches two battleships steaming off the coast of Japan. An aircraft carrier steams in the background.

From the first days of the U.S. neutrality patrols, through the disaster at Pearl Harbor, to the Japanese surrender aboard the U.S.S. *Missouri* in Tokyo Bay, the U.S. Navy played a major role in achieving the final victory. It was no easy matter. Between December 1941 and August 1945, U.S. shipyards built 8 battleships, 140 aircraft carriers of various types, more than 40 cruisers, about 350 destroyers, 575 smaller destroyer escorts, 278 submarines, and thousands and thousands of smaller ships and landing craft.

The Navy had driven the U-boats out of the Atlantic, landed troops all over the world, and choked Japan in the Pacific. In short, it had fought a hard, conventional, all-out war and won. It had preserved American society and saved democracy in Europe.

As the guns grew silent and the ships came home flying their battle flags for the last time, Navymen knew they had done their job—and done it well. However, not many of them realized that the world had changed and that a new era had dawned. This unfamiliar period—the Cold War—would be a different kind of world. But, as we shall see, the Navy would remain an important, vital part of that new era.

This was the admiral's inspection aboard an aircraft carrier. Each flattop had a crew of up to 3500 officers and enlisted men, all working together in an effective team. Today's carriers have crews of up to 6000—truly floating cities.

Landing planes aboard a U.S. aircraft carrier in the Pacific during World War II. These ships revolutionized naval warfare. Several "sea" battles were fought entirely by carrier planes with the opposing ships never seeing each other.

Some of the hundreds of U.S. ships of every size and shape mothballed after World War II for possible future use. Many "experts" were certain that these ships never again would be needed in the atomic era.

# THE 4 COLD WAR ERA

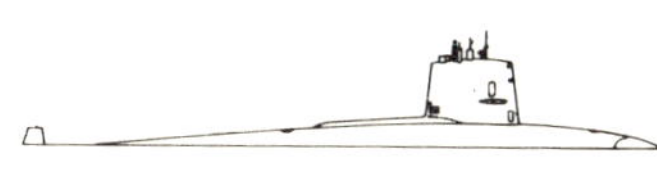

*They that can give up essential liberty to obtain a little temporary safety deserve neither liberty nor safety.*

—BENJAMIN FRANKLIN (1706–1790)

After World War II ended in 1945, millions of American men and women were taking off their uniforms and packing them away in trunks . . . to be forgotten forever. The U.S. Navy, then the world's strongest fleet, was severely cut back in size. Construction was halted on many warships, hundreds more were placed in mothballs, and thousands of bluejackets became civilians again. Most Americans thought that peace had been restored to the world and that the nation no longer needed a large navy.

But even before the war was over, there were indications that the United States might be entering a period of conflict with Soviet Russia. After the victory over Germany, Communism began pushing out from Russia's borders into Eastern Europe and Asia. In a March 1946 speech in Fulton, Missouri, former British Prime

Minister Winston Churchill said that "an iron curtain has descended across the Continent." A new, nonshooting war had emerged in the world between two differing ideas; the clash of Western democracy and Soviet Communism became the political battle of the Cold War era.

## THE TRUMAN YEARS

One of the first showdowns of the new era was over Berlin, the former capital of Germany. The city had been divided into four areas, with Britain, France, the United States, and the Soviet Union each occupying one sector. Because Berlin was situated a hundred miles inside East Germany, the Western Allies had to transport food for the population over Communist-controlled rails and highways. In 1948 Russia cut the traffic between West Berlin and West Germany, and the city was left with shortages of food and coal.

The Soviets hoped to choke the Allies and force them out of Berlin. President Harry Truman met this challenge with a massive U.S. and British airlift of material. The aircraft flying these missions of mercy in 1948 and 1949 got the fuel for their engines and much of the cargo by sea. Russia soon realized that the United States had, in effect, created a giant pipeline from the United States, Great Britain, and France to West Berlin. This meant that the Berliners couldn't be defeated through starvation, so the Soviets eventually reopened the roads and rail lines.

Then, one year later, in June 1950, Communist North Korean troops invaded the Republic of Korea to the south. Again President Truman met the challenge head on by committing the United States to driving the Red forces back.

In the early days of the war, the Navy's aircraft carriers were important backstops for the United Nations forces on the ground. Navy planes flying from the decks of the flattops were the only friendly aircraft close to the fighting because the invading North Koreans captured many of the airfields in the south. American GIs and South Korean soldiers depended on U.S. planes

In response to the Cold War confrontation between the Soviet Union and the Western Allies, the U.S. Navy began regular operations of warships in the Mediterranean, which marked the beginning of the U.S. Sixth Fleet. Here the carrier *Midway* steams through a gale off Sicily early in 1949.

to smash Communist tanks, trucks, and personnel. Taking off from carriers in the Sea of Japan, the Navy's aircraft bombed rail lines, highways, and bridges to cut the flow of supplies to the North Korean troops.

Shortly after the war began, the Navy carried out a brilliant amphibious landing at Inchon. The operation, planned by General Douglas MacArthur, enabled U.S. Army troops to slip behind enemy lines and retake a great deal of territory.

Throughout the bitter fighting in Korea, U.S. Navy battleships, cruisers, and destroyers shelled Communist troops ashore. The Navy also assisted the war-torn nation by carrying Korean civilians away from the fighting, and this saved many innocent lives. In addition, Navy minesweepers cleaned out harbors mined by the enemy.

Merchant marine and naval transport ships were very important to the whole war effort. With about a half million U.S. and U.N. troops fighting on the land, the United States had to

Jets and windmills. The Korean War introduced two new aerial weapons to naval warfare: the jet-propelled aircraft, such as this F9F Panther about to be launched from a carrier, and the helicopter, such as the H03S hovering nearby to pick up the pilot in case of an accident.

carry most of the soldiers, and 99 percent of the food, munitions, fuel, guns, and equipment across the Pacific during the three years of the war.

As in World War II, the United States needed to use an ocean highway to keep men and supplies moving from the home front to the battlefield thousands of miles away. Therefore, while the Navy could not win the Korean War alone, the war would have been lost without the Navy.

## LATER CRISES

After the Korean armistice of 1953, the United States had to preserve the independence of non-Communist governments around the world on several occasions. The Navy, as the seagoing arm of American policy, answered the call from Washington throughout the 1950s and 1960s.

In September 1954, for example, President Dwight

Eisenhower ordered the Navy to assist the Chinese Nationalists in defense of Quemoy Island, off the coast of Mainland China. Less than a year later, U.S. Pacific Fleet ships evacuated the Nationalist forces from Tachen Island. The presence of the fleet in those waters kept the peace between the rival Chinese governments.

President Eisenhower used the Navy again in 1958 to assist a friendly government. When Lebanon appealed for American aid to head off a possible Communist-inspired revolution, the United States landed three reinforced Marine battalions from Navy ships based in the Mediterranean.

The next President, John F. Kennedy, also used U.S. sea power in times of crisis. Soviet offensive missiles had been secretly transported to Cuba in 1962, and Washington saw them as a direct threat to the United States and other Western Hemisphere nations. (The low-level photographs which had

During two world wars, Korea, and later Vietnam, ships that carried men, supplies, and equipment to the battlefield and overseas bases played a vital role in the total war effort.

Low-flying U.S. Navy and Marine photo reconnaissance planes brought back views such as this one of the Cuban port of Mariel. It clearly showed how the Soviets were moving missiles and bombers into Cuba.

revealed the missile buildup had been taken mainly by Navy and Marine RF-8A Crusader photo aircraft.)

President Kennedy decided to place a naval quarantine on Cuba, preventing the Russians from supplying Fidel Castro. The Navy used 180 ships, including aircraft carriers and many destroyers, plus patrol aircraft to monitor Soviet merchant ships en route to the island. Some merchant ships were stopped and inspected for missiles and bombers. The Soviet leadership soon realized that the U.S. Navy controlled the waters off Cuba, and the missiles in Cuba were removed.

President Lyndon Johnson used the Navy-Marine Corps team early in his administration to achieve a political result. Civil unrest in the Dominican Republic in 1965 threatened the role of

The Navy was on the scene during the 1962 Cuban missile crisis: a destroyer (stern at far left) and a Navy patrol plane take a close look at the Soviet freighter *Volgoles* carrying missiles on her deck.

U.S. combat participation in the Vietnam War began in 1964 when this North Vietnamese PT boat and several others attacked U.S. destroyers in the Gulf of Tonkin. This photograph was taken from the U.S.S. *Maddox* as she fought off the attackers.

law in that Caribbean nation. The U.S. ambassador reported that American lives were in danger from disorders in the streets of Santo Domingo, the capital, and the President ordered a Navy task force into the area. Using helicopters from an amphibious ship, some 23,000 Marines were eventually flown into the city to preserve order and save lives. But it was in another republic, thousands of miles away, that President Johnson used sea power on a large scale.

## VIETNAM

The U.S. Navy played an important role during the Vietnam conflict, even from the beginning of the major U.S. involvement there. Starting with an August 1964 attack by North Vietnamese PT Boats against the U.S. destroyer *Maddox* in the Gulf of Tonkin, Navymen were in the thick of the fighting. They performed a number of different and difficult jobs, and their assignments and successes tell us much about the sailors' talent, vigor, and training.

Naval airplanes on aircraft carriers in the Tonkin Gulf went into action immediately. During the initial part of the war, the big carriers with nearly a hundred planes were the key strike air bases in the area. The floating airfields at "Yankee Station" in the northern waters and "Dixie Station" farther south hit the enemy's supply network, and supported Allied ground troops. Unlike the

A rare view of the U.S. Seventh Fleet operating in the South China Sea during the Vietnam War. Normally the four aircraft carriers shown in the photograph would be separated by many miles.

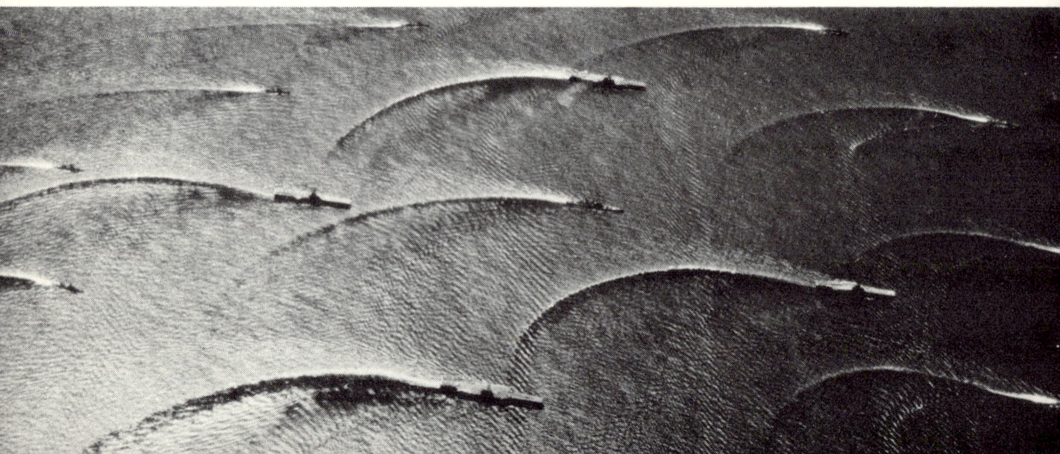

An important mission of the U.S. Navy in Vietnam was the interruption of Communist efforts to send men and arms into South Vietnam by sea. Here a U.S. Navy patrol craft approaches one of the thousands of junks in Vietnamese waters to search for hidden arms or supplies.

Vietnam saw the rebirth of the U.S. Navy's "brown water"— or river and inshore—activities. Here Navy riverine gunboats lead troop carriers up a river. Every sailor and gun are "ready for anything."

Mother and brood: twenty-five Navy riverine craft rest alongside a support ship during the Vietnam War. The U.S. Navy
operated almost eight hundred small combat craft in the
conflict; subsequently this "brown water" fleet was turned
over to the South Vietnamese Navy.

land bases in Vietnam, the aircraft carriers were free from
Vietcong ground attack, and this made them very effective,
secure, and valuable.

The fighting during the Vietnam conflict often resembled the
American Civil War. For example, Navy and Coast Guard ships
were given the hard job of stopping enemy guerrillas from slip-
ping into South Vietnam by sea. Operation Market Time, a
modern version of the Civil War blockade, was used to search
Vietnamese junks for hidden guns, ammunition, and soldiers.
Vietnam's many inland waterways served as back roads for innocent civilians and for the enemy. The Navy created a whole
new "brown water" river force to cut Vietcong traffic along the
rivers and canals, and to attack the guerrillas in their own
backyards. Using converted landing craft, special types of boats,

The world's last active battleship was the U.S.S. *New Jersey*, employed briefly as a fire-support ship during the Vietnam conflict. The giant, shown here firing against Communist forces in South Vietnam, is again mothballed in the reserve fleet along with three sister ships. All four battleships are veterans of World War II and Korean combat.

and helicopters, American sailors battled along the twisting, narrow, and shallow waterways.

Navy destroyers, cruisers, and the world's last active battleship—the U.S.S. *New Jersey*—shelled North Vietnam and Vietcong positions from the sea as part of Operation Sea Dragon. These were very useful in destroying enemy troops and supplies when they threatened U.S. and Vietnamese bases.

Previous wars had shown the importance of moving material to overseas combat areas by ship, and the Vietnam war was no exception. Although most of the fighting was on the ground and in the air, almost all of the supplies needed to support the Allies came by sea. Even small, short-range aircraft, such as helicopters, were carried on the decks of naval and merchant ships.

Navy construction specialists, called Seabees, operated

bulldozers, dump trucks, cranes, and cement mixers to construct airfields and port facilities ashore. These combination builder-sailors moved tons of earth and laid miles of concrete to create all new runways, highways, docks, warehouses, offices, schools, and depots.

One of the Navy's last missions during the war was the mining of North Vietnam's harbors in May 1972. Dropping antiship mines, U.S. Navy aircraft effectively closed the ports for a year. This bold move ordered by President Nixon shut one of the major points of entry to North Vietnam and made great problems for the Communists. (Later, after the fighting was stopped in January 1973, a U.S. Navy task force was sent to Haiphong to help with the sweeping of the mines.)

During the last years of the American combat role, the U.S. Navy assisted the Vietnamese in preparing their own navy to defend the country. The Navy transferred several oceangoing ships for offshore patrol and hundreds of coastal and river craft to the Vietnamese.

After a quarter century of Cold War and crises, today's Navy has served the nation in places all around the earth. American Presidents in the post-World War II era learned the unique advantages of having a strong Navy. They have seen that ships at sea are free from the problems of ground forces on foreign soil, that the Navy is a versatile and powerful arm of the United States, and that it can be depended upon to defend liberty at any time and anywhere.

In the next four chapters we will take a closer look at the Navy today—in the air, on the ocean, and beneath the surface, and, most importantly, at the people who serve in its ranks.

# 5 Navy Missions

*Every danger of a military character to which the United States is exposed can be best met outside her own territory—at sea.*

—REAR ADMIRAL ALFRED T. MAHAN, U.S. NAVY (1840–1914)

Armed forces exist to preserve and defend a nation and to help the country carry out its goals. This is true of the United States Armed Forces and, because of the geographical position of the United States—between the world's two great oceans—and our overseas interests, the Navy has particular importance.

Today one could argue that navies are obsolete because of airplanes, satellites, nuclear weapons, and other modern technology. Not only is that not the case, but because of this nation's *increasing* dependence upon the seas to bring in raw materials and carry out manufactured goods, and the nation's great overseas economic and political interests, oceans may be more important today than ever before in our history.

In this age the U.S. Navy has four principal missions:

— Nuclear deterrence
— Sea control
— Projection
— Presence


## Nuclear Deterrence

This means providing the nation with nuclear weapons which must survive an enemy's surprise attack against the United States. The enemy would know that even if he destroyed the nation, sufficient U.S. nuclear weapons would survive the attack to enable the United States to cause equal damage to his nation. Thus, the survivable nuclear weapons deter the enemy from launching a surprise attack. This is always a possibility whenever an aggressor thinks, for political or military reasons, that the U.S. nuclear striking forces could be destroyed in a surprise, first-strike attack.

Today the United States has three so-called "assured destruction" forces: land-based bombers, land-based missiles (ICBMs), and submarine-launched ballistic missiles (SLBMs). The SLBMs are the Polaris and Poseidon missiles at sea in the Navy's nuclear-propelled ballistic missile submarines.

These submarines and their missiles are probably the most important part of the nation's nuclear deterrence because they are the least vulnerable to a surprise enemy attack or are the most "survivable" of the nation's assured destruction forces. Submarines at sea are always moving in the ocean depths, where they are difficult to locate, follow, and attack. In contrast, the locations of land-based bombers and ICBMs are always known; an enemy can plan to make a surprise attack tomorrow at noon, or next week or next month. He cannot plan an attack against Polaris or Poseidon missile submarines because he simply does not know where they will be even a few hours from now.

Also, the Soviet Union has built up a superiority of nuclear

Navy Polaris and Poseidon ballistic missile submarines are the nation's most survivable nuclear deterrent. These submarines, continually cruising in the ocean depths, ensure that a nation attacking the United States with nuclear weapons will itself be destroyed. This is an artist's concept of a ballistic missile submarine firing from beneath the Arctic ice.

strike weapons during the past few years which could attack U.S. land-based bombers and missiles. However, the Soviets have not developed the large and sophisticated antisubmarine forces that would be required to attack U.S. missile submarines.

Today the U.S. Navy's missile submarine forces consist of forty-one undersea craft armed with long-range Polaris and Poseidon missiles; each missile carries a nuclear warhead with several times the explosive force of all the conventional bombs and missiles ever exploded. In addition to these forty-one submarines, more than half of which are at sea at any given time, the Navy is developing a new type of missile-armed submarine under the Trident program. The Polaris/Poseidon and Trident submarines are described in more detail in Chapter 8.

NAVY
507
507

The Navy's larger aircraft carriers operate jet-propelled bombers that can deliver nuclear weapons against an aggressor's military bases or cities. However, this nuclear strike capability is considered a secondary role for aircraft carriers; they are primarily regarded as sea control forces.

## Sea Control

This is the concept of being able to use an ocean area to support a nation's interests, and denying the seas to the ships of other nations. For example, if we have to send ship convoys to Europe, sea control forces would escort the merchant ships, protecting them from enemy aircraft, submarines, and surface warships.

Sea control means the use of naval forces, including the destroyer, the P-3 Orion patrol plane, and the SH-3 antisubmarine helicopter shown here, to ensure that friendly forces can use the seas in safety.

Also, sea control could mean going into an area where an enemy is trying to move men or supplies by ship and stopping them, by sinking or capturing their ships, or by planting mines on the ocean floor which could stop their ships without a direct confrontation with U.S. warships.

Sea control is perhaps the oldest and most traditional naval mission. Before this century, one had only to counter surface ships to perform this job; the development of submarines and aircraft in the early 1900s made this mission more difficult. Today, the Soviet Navy possesses the world's largest submarine and surface fleets, plus a large naval air arm. Thus, sea control missions require that the U.S. Navy have the weapons to counter this "three-dimensional" threat because many of the Soviet submarines, surface ships, and aircraft carry very sophisticated missiles for attacking ships.

Potential sea control missions in non-Soviet conflicts require that the U.S. Navy also be capable of combating relatively primitive sea forces. For example, during the Vietnam War, the U.S. Navy had the mission of stopping the Communist infiltration of men and arms into South Vietnam by sea in junks and other small craft.

The Navy's sea control forces thus include ships for countering both sophisticated and simple threats at sea. These forces include aircraft carriers and their planes; various surface warships, such as cruisers, destroyers, frigates, and escort ships; small mine, patrol, and missile ships and boats; submarines armed with torpedoes; and land-based patrol aircraft. More detailed descriptions of these ships, submarines, and aircraft are given later in this book.

A large portion of these sea control forces are at sea at any given time. The U.S. Sixth Fleet in the Mediterranean normally has two aircraft carriers and perhaps thirty other ships and submarines, plus land-based patrol aircraft. Similarly, the Seventh Fleet in the Western Pacific has about three aircraft carriers, perhaps forty other ships, and numerous land-based aircraft. Additional ships, submarines, and aircraft operate in other areas in which the United States has interests.

Many of these naval forces also can undertake projection and presence missions in support of national interests.


**Projection**

The extension of U.S. power into land areas from the sea is called projection. In the post-World War II era, naval forces were used to project air and land striking forces ashore in the Korean War, the 1958 Lebanese landings, and the Vietnam War.

The primary means of projection by naval forces consists of aircraft based aboard carriers and amphibious ships, such as LSTs, which carry Marines and their weapons. Other ships are required to support the carriers and amphibious ships, provide gunfire against shore targets (until the Marines have their own artillery set up ashore), sweep mines, and survey beaches.

Aircraft carriers are a primary means of projecting U.S. power into overseas areas. Here a Navy jet attack plane takes off from the nuclear-propelled carrier *Enterprise*.

The Sixth Fleet in the Mediterranean and Seventh Fleet in the Western Pacific have projection capabilities in their carriers, plus an amphibious task force of about five ships in each fleet, with a Marine battalion landing team of about 2000 men embarked. Additional carriers and Navy-Marine amphibious forces can be sent to sea and rushed to crisis areas when required. The self-sufficiency of naval forces and their ability to replenish at sea permits these mobile projection units to be moved into a crisis area and kept there or be employed as needed. They can be used without establishing forward bases on foreign territory and without the permission of other nations. In fact, these naval forces can be sent into an area in some situations without even the knowledge of other countries. In addition, the U.S. merchant marine projects American influence into overseas areas, and can be regarded as a "projection force."

Naval forces are the most versatile instruments of national

Navy jet attack planes and fighters rest on the deck of the carrier *Ranger* while other carriers maneuver in the background during operations in the South China Sea.

Marine-carrying landing craft head for shore during an amphibious exercise. Note the helicopter at far left taking off from an "amphib" ship.

policy in that their mobility, flexibility, and sustained readiness also permit them to serve as presence forces.


## Presence

This is a means of demonstrating to another nation that the United States has an interest in an area—for political, military, or economic reasons. The ability to move naval forces rapidly into an area can influence another nation, such as the Soviet Union, to stay out of the locale because of the risk of a direct confrontation.

In this respect, U.S. ships in another nation's waters—with the permission of that country—demonstrate the friendship and support of the United States. For example, U.S. ships in the Formosa Straits are believed to have deterred an attack by Mainland China against the Nationalist regime on Taiwan, and American ships off the coast of Cyprus have deterred Turkish

On a day-to-day basis, the U.S. Navy is "flying the flag" in foreign ports, such as the destroyer *Harwood* shown here entering Valletta Harbor, Malta, a strategic crossroads of the Mediterranean.                    *Anthony & Joseph Pavia*

military forces from interfering in the affairs of that troubled island.

Naval presence is a time-proven naval activity and is also practiced today by the Soviet Union in its deployment of large numbers of ships overseas. In some places in the world, U.S. presence forces assure allies and neutral nations that the Soviet Union will not interfere in their domestic affairs.

The various naval forces required to carry out the missions described above include complex ships, submarines, and aircraft which incorporate some of the world's most advanced technology: nuclear propulsion, supersonic aircraft, electronics, lasers, guided missiles, and advanced marine systems, such as surface effect ships, air cushion vehicles, hydrofoils, gas turbines, deep-diving submarines, and even navigation and communication satellites.

## Support

Support for these naval forces requires, in turn, sophisticated support ships or "tenders," repair ships, oilers, ammunition ships, and seagoing tugs. Ashore there are bases, depots, shipyards,

A destroyer tender or depot ship provides services to four destroyers moored alongside. The tender is a floating repair/-supply ship especially fitted to service the needs of destroyers and their sailors.

research and development facilities, communications and intelligence centers, weather stations, schools and training centers, and the almost innumerable other activities for supporting a modern navy. Together, the nuclear deterrence, sea control, projection, presence, and support forces make today's Navy the strongest fleet ever to sail the seas.

Figure 1 below shows some of the equipment used to carry out these missions. In the following chapters, we will discuss this equipment in terms of the air, surface, and submarine forces of today's Navy.

**Figure 1. Navy Missions and Equipment**

| Mission | Forces |
| --- | --- |
| Nuclear Deterrence | — Polaris/Poseidon missile submarines<br>— submarine support ships or tenders<br>— aircraft carriers |
| Sea Control | — surface warships, such as cruisers, destroyers, frigates, escort ships<br>— torpedo-carrying submarines<br>— small mine, patrol, and missile ships and boats<br>— aircraft carriers and their planes<br>— land-based patrol planes |
| Projection | — aircraft carriers and their planes<br>— amphibious ships, such as LSTs<br>— Marine landing forces<br>— surface warships |
| Presence | — most naval forces |
| Support | — auxiliary ships, such as tenders, repair ships, oilers, ammunition ships, tugs, etc.<br>— land bases, depots, shipyards<br>— research and experimental facilities and ships<br>— communications, intelligence, and weather forecasting activities, etc.<br>— schools and training centers |

The modern aircraft carrier is a floating airfield and city some three city blocks long. The nuclear-propelled *Enterprise*, shown here, carries some one hundred aircraft, is manned by some 5500 officers and enlisted men, and can move at more than thirty miles an hour.

# The
# 6 Navy Today:
# Air

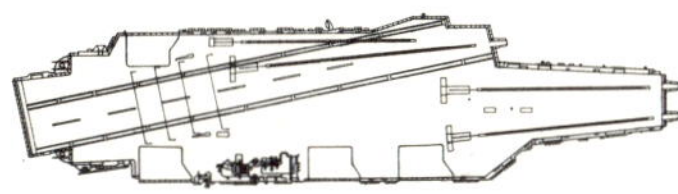

*This is the most important landing of a bird since the dove flew back to the Ark.*



—COMMENT AFTER THE FIRST
AIRPLANE LANDED ON A SHIP
ON JANUARY 18, 1911

Airplanes are an important part of everyday life—in both war and peace. Indeed, probably no other "weapon" has had as great an impact on military activities in this century as has the airplane.

This was especially true in naval warfare. Early airplanes were used to scout out the enemy's fleet or to lay smoke screens to hide friendly ships. Later some aircraft began shooting at other planes, and "fighters" came into the fleet; and soon, planes began carrying bombs and, then, torpedoes to attack enemy ships.

Early naval airplanes were seaplanes, with floats or pontoons which were used for taking off and landing on the surface of the sea. After that, sailors found a way of shooting or "catapulting"

Man's first airplane flight from a ship came in late 1910 when Eugene Ely, a civilian, took off from a U.S. Navy cruiser. This historic Navy experiment was followed by the development of ship-based aviation, including aircraft carriers, by the U.S., British, French, and Japanese navies.

seaplanes off small tracks on ships (the planes still had to land on the sea to be picked up by a crane).

Then Navymen began flying aircraft from short runways and decks built onto warships. These planes had wheels which were less awkward than floats and made the planes faster and more maneuverable. The concept of the aircraft carrier came from the early experiments with wheeled planes aboard ships in World War I.

After 20 years of development between the world wars, the aircraft carrier emerged as probably the most important surface ship of World War II. In the Pacific, U.S. and Japanese carriers or flattops battled for control of the seas, struck out at shore bases, and provided air support for Marine assaults.

During the war in the Atlantic and Mediterranean, U.S. and British carriers fought German submarines (U-boats) and provided air support for the invasions of North Africa. However, the availability of land bases, first in Great Britain and then in North Africa and Europe, reduced the need for carriers.

After World War II, the aircraft carrier remained the principal surface ship of the U..S., British, and French navies (and recently the Soviets have built aircraft carriers).

Today's aircraft carrier is a moving island. It is some 1000 feet long, or the length of three football fields; has a crew of 5000 men; and can operate 70 to 100 modern jet aircraft, about the number operated on two or three military airfields ashore.

The modern carrier contains aviation fuel and weapons for its airplanes; computers, radios, televisions, radars, and other electronic equipment; facilities for the command and control of aircraft and ships; and other components of both an effective air base and a fighting ship.

The distinctive feature of the carrier is its flight deck. There are two sections of the deck—an angled or canted landing area and

Navymen service aircraft on the flight deck of the carrier *Enterprise.* A flattop can provide the same maintenance and support to aircraft as the most complex airfields ashore, and sometimes at less cost.

This A-4 Skyhawk attack plane is about to touch down on the deck of the aircraft carrier *Hancock*. An instant later the plane's extended tail hook (seen between the wheels) engaged one of the ship's steel arresting wires that slow the plane to a safe stop after a deck run of some three hundred feet.

a straight parking/takeoff zone. Jet aircraft come in to the angled deck at speeds approaching a hundred miles per hour (a few fighters come in even faster).

As the plane streaks over the angled deck, it lowers a hook to catch one of three or four steel wires stretched across the deck. The hook engages, and the plane is brought to a safe halt after rolling only about three hundred feet. If the airplane misses the wires, it can accelerate, take off from the clear angled deck, and come around for another landing.

After coming to a full stop, the airplane retracts the hook, which releases the wire, and then moves directly into a parking area or is carried by a large elevator down below the flight deck to the hangar deck.

Radar and other electronic equipment permit aircraft to land at night and in bad weather if necessary. The men aboard the carrier can land the plane, with the pilot keeping his hands off the stick, by watching and controlling the airplane with electronic devices.

To launch aircraft, the modern carrier has two to four catapults each with a track covering a steam-powered piston. Aircraft are attached to the piston; then, upon signal, the piston shoots forward, accelerating the planes from a dead stop to over 180 miles per hour in just 310 feet. Each catapult can launch aircraft once every sixty seconds, meaning that the carrier can shoot planes into the air at the rate of four per minute when using all four "cats."

## THE PLANES AND MEN

Navy carrier-based planes include the top military airplanes flying in the world today: F-4 Phantom fighters and the new F-14 Tomcat, both supersonic planes that carry long-range missiles for attacking enemy planes or ground targets; A-7 Corsair and A-6 Intruder bombers, planes that can carry missiles and bombs for attacking ships or shore targets, with the Intruder having an unequaled all-weather/night performance; S-2 Tracker, and S-3 Viking aircraft, and SH-3 Sea King helicopters that can seek out and attack enemy submarines.

The crews of these planes range from one man in the A-7 Corsair to five in the E-2 Hawkeye. The E-2 is capable of staying aloft for several hours, with its radar and other electronics providing a full picture of the air and surface for hundreds of miles in all directions. There are two pilots and three equipment operators. All are highly trained and dedicated men. (More about these men and other Navymen in Chapter 8.)

These planes and their pilots are the key parts of the naval air team. The carrier from which they fly is a floating base which can move over six hundred miles per day at cruising speeds, remain at sea for months at a time, and move into a crisis area when needed without putting American troops ashore or building up bases on foreign territory.

Once the carrier is on the scene—a matter of hours or a few days at most—the planes are ready to go; they can fly over a

The Navy's newest carrier plane is the F-14 Tomcat fighter. The plane's wings change angle automatically as the Tomcat maneuvers, providing the most efficient wing "sweep." The two-man, twin-jet fighter has missiles that can intercept enemy planes more than fifty miles away.

*Grumman Aerospace Corp.*

A large SH-3 Sea King helicopter hovers over a U.S. submarine during exercises. Just below the SH-3 is a sonar that can be lowered into the water by cable, one of the helicopter's several submarine detection devices.

A carrier air wing has a variety of aircraft for performing different missions. One of the more unusual is the "saucer" topped E-2 Hawkeye airborne early-warning aircraft. The large radar mounted on top of the plane and other equipment permit an E-2 to search for hundreds of miles to detect hostile aircraft and ships, and to direct friendly aircraft.

country to support a friendly government or fly combat missions as directed by the President. The versatility of the seventy to one hundred planes on the carrier, and the carrier's mobility and other features, make this a most important part of the Navy's projection and sea control forces.

In addition to aircraft carriers, naval aviation also goes to sea in destroyers, frigates, and escort ships. These ships—which are primarily for antiair and antisubmarine warfare—can carry helicopters such as the SH-2D Sea Sprite. The three-man Sea Sprite can be used to hunt and attack submarines, or can serve as an airborne radar platform to provide surveillance of sea areas and warning of enemy attack.

Also, a new type of ship is being developed called the "sea control ship." This ship resembles a small aircraft carrier but lacks the big ship's advanced aircraft and speed. The sea control ship will have about fifteen helicopters for hunting submarines

Vertical/short take-off and landing (VSTOL) aircraft, such as the AV-8 Harrier, can operate from short runways or small ships. They can take off and land straight up and down, but are more efficient with short takeoff runs.

and surveillance, plus about three of the so-called VSTOL planes (for vertical/short take-off and landing). A VSTOL aircraft is a fixed-wing plane, but it can go straight up or down like a helicopter. It is most efficient when it has a short takeoff run, perhaps three hundred or four hundred feet, although it has a smaller payload (of fuel and weapons) than conventional aircraft.

Thus naval aircraft—and aviators—are found at sea on a number of ships. (Also, helicopters fly supplies, mail, movies, and people—but more about those whirlybirds in a later chapter.)

## LAND-BASED NAVAL AVIATION

Naval aircraft also fly from land bases because of the limited number of aircraft carriers available and the ability of some

The U.S. Navy operates a large number of land-based combat, training, and transport aircraft. These are P-3 Orion, long-range aircraft used for maritime reconnaissance and antisubmarine missions.

planes to do the job from ashore. The principal land-based naval aircraft is the patrol plane.

The U.S. Navy's patrol aircraft is the P-3 Orion, a large airplane with four prop-jet engines and a crew of ten. Patrol planes fly long missions seeking out submarines (which they can attack with torpedoes, depth charge bombs, and rockets) and surface ships. These are vital when the Navy wants to know "what's out there." These planes operate regularly from coastal bases in the United States as well as airfields overseas.

Also based ashore are the Navy's training planes, a small

fleet of cargo planes that fly high priority parts and equipment to the fleet (with some of these aircraft able to land on carriers), and some special-mission planes for oceanographic survey work, electronic reconnaissance, hurricane reporting, flying supplies to scientists at the South Pole, and similar jobs.

With some 6700 airplanes in service today, the U.S. Navy is an "air navy"; this is the largest air organization in the world except for the U.S. Air Force and the Soviet Air Force. These aircraft, the men who fly them, and the ships and shore bases from which they fly are vital to the U.S. Navy.

A P-3 Orion heads to sea on a surveillance mission. The Orion has several different submarine detection devices, computers for analyzing data received from these devices, and can carry bombs, torpedoes, rockets, and missiles for attacking hostile ships or submarines.

**Figure 2. NAVY MISSIONS AND AIR FORCES**

| Mission | Forces |
|---|---|
| Nuclear Deterrence | — Polaris/Poseidon mission submarines<br>— submarine support ships or tenders<br>— AIRCRAFT CARRIERS |
| Sea Control | — surface warships, such as cruisers, destroyers, frigates, escort ships<br>— torpedo-carrying submarines<br>— small mine, patrol, and missile ships and boats<br>— AIRCRAFT CARRIERS AND THEIR PLANES<br>— LAND-BASED PATROL PLANES |
| Projection | — AIRCRAFT CARRIERS AND THEIR PLANES<br>— amphibious ships, such as LSTs<br>— Marine landing forces<br>— surface warships |
| Presence | — most naval forces |
| Support | — auxiliary ships, such as tenders, repair ships, oilers, ammunition ships, tugs, etc.<br>— land bases, depots, shipyards<br>— research and experimental facilities and ships<br>— communications, intelligence, and weather forecasting activities, etc.<br>— schools and training centers |

# The
# 7 Navy Today:
# Surface

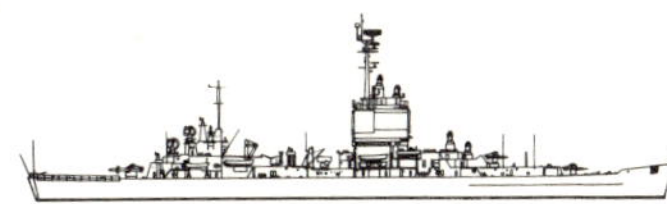

*They that go down to the sea in ships, that do business in great waters.*

—PSALM 107:23

The earliest "naval" ships were simply fishing or trading ships pressed into service by a local prince or lord. As ships got larger and went farther from shore, they were sometimes attacked by pirates. The princes and lords soon needed armed ships to protect their commercial vessels, and this marked the beginnings of navies.

Just a few soldiers were all that those early ships needed to turn them into "warships." Then, special platforms were installed from which the soldiers could fight. Some countries later added rows of oars to give the warships more speed over sailing craft. Next, after the invention of gunpowder, cannon were installed.

Still, most "warships" before the 1800s were basically merchant ships with some refinements, as seen in the frigates and

A number of developments during the 1800s led to the "modern" warships built in the 1900s. The battleship *Kearsarge*, completed in 1900, was one of the first "modern" combat ships. She was one of the U.S. battleships that sailed around the world in 1907-1908 to "show the flag" and was distinguished as the only American battleship without a state name, although she did honor New Hampshire, being named for a mountain in that state.

ships of the line of the eighteenth century. In the nineteenth century the development of advanced (rifled) guns, underwater torpedoes, armor, steam propulsion, and screw propellers led to the specialized surface ship designs of today.

Battleships were the big-gun, heavily armored warships of the early 1900s. Development of the torpedo, an "underwater missile" that could be fired against a battleship from a distance of thousands of yards, led to the creation of torpedo boats. Then, to defend battleships from torpedo craft, the torpedo boat destroyer—or, more simply, "destroyer"—evolved.

Cruisers, ships with medium-sized guns and some armor, were built to cruise in distant waters to protect a nation's interests.

Modern Navy ships employ electronics for detecting and tracking ships, aircraft, and submarines; for guiding weapons, navigation, communications, and exchanging data. The sailor in foreground is operating a tactical data system that converts radar signals and other information to a "picture" of his ship's operating area.

Then, landing ships were developed to put troops ashore in order to support those interests. Again, the first ships of this type were simply merchant ships with spaces for troops and horses. Later, ramps were designed so that the troops and horses could march off; and, finally, specialized ships were built with large bow

Missiles are an inherent component of naval forces. Surface ships carry missiles for use against hostile ships, aircraft, and submarines. Here a Terrier antiaircraft missile slides onto a launcher on the nuclear-propelled frigate *Bainbridge*.

A Terrier antiaircraft missile streaks skyward from a guided missile frigate. These surface ships provide close-in air defense for carrier forces. They also have an antisubmarine capability.

Nuclear warheads provide Navy missiles with a "tactical" atomic capability. An antisubmarine rocket fired by the destroyer *Agerholm* (foreground) created this underwater nuclear blast.

doors and ramps enabling troops and tanks to unload directly onto the beach.

The development of modern weapons, especially aircraft and guided missiles, and the advent of shipboard electronics, such as radar and sonar, led to the creation of new ship types and a blurring of some of the old distinctions. Also, some technological developments, such as nuclear reactors, gas turbines, hydrofoils (ships that ride on underwater "wings"), and surface effect ships (ships that ride on a cushion of air), further confuse the surface ship picture.

A quick look at today's U.S. Navy shows many surface ships, the principal types of which are:

Six guided missile cruisers—the largest surface warships except for aircraft carriers.* All of these ships are missile-armed and loaded with electronic equipment. They serve as defensive ships for

---

*Aircraft carriers are discussed separately in the preceding chapter.

carriers, and carry admirals who serve as task force commanders. All of these ships but one were built in World War II and have been extensively modernized; the *Long Beach,* completed in 1961, was the world's first nuclear-propelled surface warship.

Thirty guided missile frigates—sleek and fast warships, almost as large as cruisers. The frigates "screen" carriers and are armed with antiaircraft and antisubmarine missiles. Two frigates, the *Bainbridge* and the *Truxtun,* are nuclear-powered, and several more of this type are being built.

One hundred destroyers—the workhorses of the fleet, armed with guns and antisubmarine weapons, with a few carrying antiaircraft missiles. These ships are capable of high speeds (over thirty knots) so they can operate with carriers as well as provide gunfire support for amphibious landings, carry out independent reconnaissance missions, etc. A new group of thirty large destroyers of the *Spruance* class is under construction.

Sixty-five ocean escorts—formally called destroyer escorts. These ships defend amphibious and merchant ships crossing the ocean. They have weapons for fighting aircraft and submarines, and a few are armed with antiaircraft missiles.

Because of the importance of bringing raw materials into the United States, especially petroleum from the Middle East, the Navy is planning to construct an additional fifty escort-type ships. These ships are currently termed patrol frigates (PF), and will have an antiaircraft and antiship missile capability.

Fifteen missile boats and gunboats—small, high-speed craft for coastal operations. These craft protect coastal areas from the enemy. They are the successors to the famed PT boats of World War II. Several hydrofoil craft have been evaluated by the Navy and a series of thirty missile-armed hydrofoil boats is being built.

Sixty-five amphibious ships—this is the Navy's half of the famed Navy-Marine amphibious team (see Chapter 10). These ships carry Marines, their tanks, guns, and equipment, plus landing craft and amphibious tractors. Amphibious ships, some of which are assigned to each fleet, can provide a quick-reaction capability by means of one or two thousand highly capable Marines.

Sleek-looking ships such as this ocean escort are charged with the primary mission of protecting convoys of merchant ships from enemy submarines in time of war.

A sign of the times: the *Tucumcari,* one of the U.S. Navy's experimental hydrofoil gunboats, is being followed into service by thirty missile-armed hydrofoil boats. Small combat craft such as the *Tucumcari* are being built by several navies.

A wood-hulled ocean minesweeper at sea. The U.S. Navy is phasing out these craft in favor of helicopter minesweepers. The white, torpedo-like objects are floats for trailing mine-sweeping devices.

Ten minesweepers—small ships that can detect and sweep or destroy advanced antiship mines. A large number of these small ships have been phased out of the U.S. Navy because ship-launched helicopters can better perform the role of countering mines in most situations.

One hundred thirty auxiliary and support ships—these non-warships come in all sizes, shapes, and capabilities, essentially for the purpose of supporting warships. There are oilers to provide various petroleum products; store ships to carry refrigerated foods; repair ships to fix anything from steel plate to intricate electronic equipment; ammunition ships to carry bullets, bombs, rockets, and missiles; tenders to provide services, spare parts, and special support to submarines and destroyers; oceanographic and survey ships

The U.S. Navy's great mobility and flexibility is due in large part to underway replenishment (UNREP) forces, including such ships as the large oiler-ammunition ship *Sacramento*, shown here fueling an aircraft carrier. Note the helicopter approaching the carrier which is delivering supplies transferred by *vertical* replenishment (VERTREP) from the *Sacramento*.

to investigate and explore the oceans and the ocean floor; salvage ships and tugs; and, finally, a score of experimental ships employed in developing the techniques and equipment of tomorrow's Navy.

Together with aircraft carriers, submarines, and naval aviation, these ships carry out the various missions assigned to Zumwalt's Navy. Figure 3 illustrates these missions, and the roles assigned to surface ships.

Today, in several areas of the world, task forces consisting of various types of ships support the interests of the United States. These ships, the planes on their decks, and crews who man them are in many respects the most essential aspect of America's future role as a major world economic, political, and military power.

**Figure 3. NAVY MISSIONS AND SURFACE SHIPS**

| *Mission* | *Forces* |
| --- | --- |
| Nuclear Deterrence | — Polaris/Poseidon missile submarines<br>— submarine support ships or tenders<br>— aircraft carriers |
| Sea Control | — SURFACE WARSHIPS, SUCH AS CRUISERS, DESTROYERS, FRIGATES, ESCORT SHIPS<br>— torpedo-carrying submarines<br>— SMALL MINE, PATROL, AND MISSILE SHIPS AND BOATS<br>— aircraft carriers and their planes<br>— land-based patrol planes |
| Projection | — aircraft carriers and their planes<br>— AMPHIBIOUS SHIPS, SUCH AS LSTs<br>— Marine landing forces<br>— SURFACE WARSHIPS |
| Presence | — MOST NAVAL FORCES |
| Support | — AUXILIARY SHIPS, SUCH AS TENDERS, REPAIR SHIPS, OILERS, AMMUNITION SHIPS, TUGS, ETC.<br>— land bases, depots, shipyards<br>— research and experimental facilities and ships<br>— communications, intelligence, and weather forecasting activities, etc.<br>— schools and training centers |

# The
# 8 Navy Today:
# Submarines

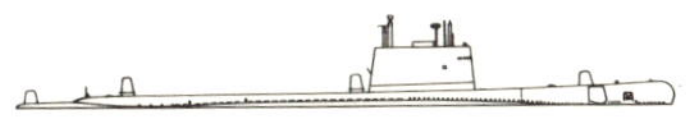

*He goes on a great voyage that goes to the bottom of the sea.*

—GEORGE HERBERT,
POET (1593–1633)

Since man first saw the sea he has sought to go below its surface, to reap the harvest of food, precious stones, and other wealth of the oceans, and also to fight his enemies. Ancient literature tells of divers who swam under enemy ships to cut their anchor ropes or cut holes in their bottoms.

Later, man began making "submarines," or craft to carry them into the depths. Probably the first use of a submarine in war occurred during the American Revolution when David Bushnell, a student at Yale, built a wooden craft shaped like a walnut standing on end. Inside, a man turned cranks to propel the craft and submerge it by an ingenious system of screws. Dubbed "Turtle," the submersible attempted to blow up the British ship *Eagle* in New

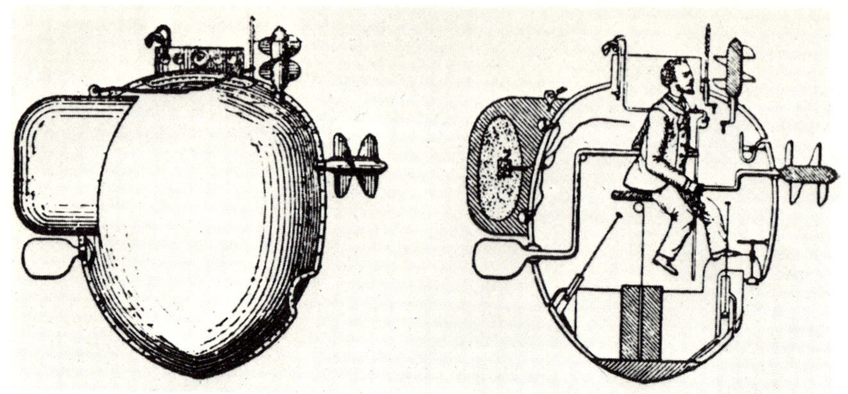

David Bushnell's submersible "Turtle" attacked British ships
in the American Revolution and the War of 1812, as Americans
pioneer submarine developments.

York Harbor on July 12, 1776, using a charge of gunpowder that
was to be attached to the warship's hull. The plan failed, but the
"Turtle" escaped; a similar Bushnell effort during the War of 1812
also failed.

Subsequently many brave Americans—among them artist-in-
ventor Robert Fulton—developed undersea craft. Practical submarines became possible in the late 1800s with the development of
electric motors to power the boats underwater without either
producing excessive heat or requiring oxygen for combustion.
These early submarines ran on gasoline engines (later diesels) that
would propel them on the surface and recharge batteries for the
electric motors.

Thus, at the start of the twentieth century, the U.S. Navy
accepted its first official submarine, the 54-foot-long *Holland*,
named for the Irish immigrant schoolteacher who built the craft for
the Navy. Submarines designed by John Holland and those of his
contemporary, Simon Lake, were soon appearing in the navies of
Britain, Japan, and tsarist Russia.

These early undersea craft (called "boats") were small,
generally damp, and smelled of oil. German historian Harold Busch
would later write, "To those who have never been to sea in a
submarine, it is hard indeed to convey an adequate idea of what it

The U.S. Navy's first "official" submarine was the *Holland*, a 54-foot craft completed in 1900. Here the submarine sits in a dockyard as modifications are made.

The *Holland* was used to train the U.S. Navy's first submariners. Similar undersea craft also served in the British, Russian, and Japanese navies. Here the *Holland* shares a berth with the Russian battleship *Retvizan*.

During World War II, American submarines severed the pipeline of merchant ships from the oil-rich Dutch East Indies to Japan, a key reason for Japan going to war. The *Barb* was typical of the U.S. submarines of the period and was one of the top-scoring subs.

means to live, sometimes for months on end, in a narrow tubular space amid foul air and universal damp."

Submarine development in the U.S. Navy continued. U.S. undersea craft played no significant part in World War I, but in the Second World War, American-manned submarines had a vital role in the Pacific. They sank 55 percent of the Japanese merchant fleet, as well as numerous warships, and contributed greatly to destruction of the Japanese empire. This achievement was accomplished by a force that consisted of *less than 2 percent* of the U.S. Navy's personnel. (Out of 288 U.S. submarines that fought in the Pacific War, fifty-two, or almost one out of every five, were lost.)

After the war the U.S. Navy had relatively little interest in submarines because the nation's potential enemy, the Soviet Union, had neither a naval nor merchant fleet. Still, there was some interest in adapting German-developed technologies to submarines, and existing submarines were modernized and a few new

The nuclear submarine *Nautilus* was the world's first true submarine, capable of operating underwater totally independent of the earth's atmosphere for sustained periods. This is a rare photograph of the *Nautilus* under way at a high speed on the surface.

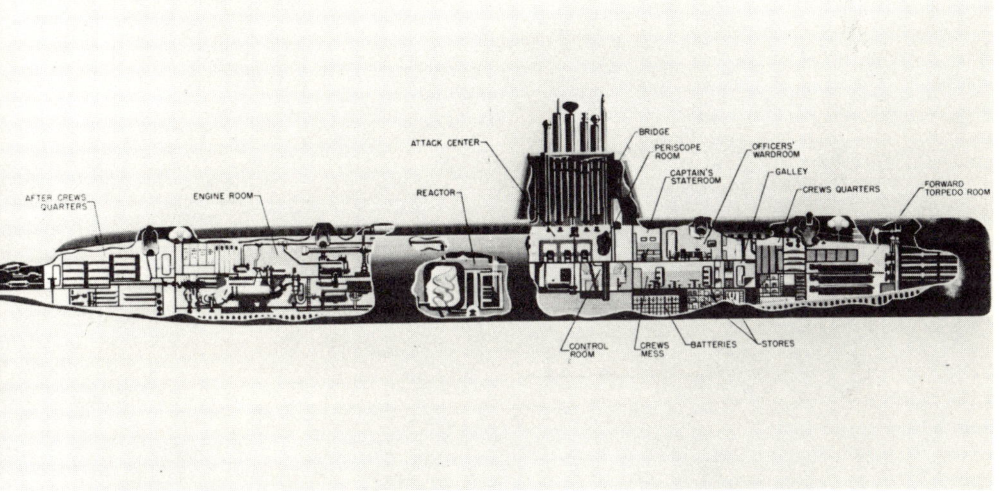


A cutaway view of the nuclear-propelled submarine *Nautilus*.

The nuclear-powered submarine *Nautilus* enters New York Harbor after her 1958 transpolar voyage under the Arctic ice pack.

"boats" were built. But in this period the U.S. Navy undertook history's most important submarine development. Previously, submarines were really surface craft which could submerge for only a few hours before having to return to the surface to recharge their batteries. For example, the most-advanced U.S. diesel submarines could only operate underwater at eighteen knots for a few hours before having to recharge batteries by running their diesel engines, either on the surface or with the help of a short tube to the surface (snorkel) which sucked in air for combustion in the diesel.

The development of nuclear propulsion in the 1950s freed submarines from having to return to the surface and gave them an underwater endurance measured in years rather than minutes. Under the direction of then-Captain Hyman G. Rickover, the Navy and several industrial firms developed the nuclear-powered submarine, man's first application of nuclear energy to nonbomb purposes.

President Harry Truman spoke at the start of construction of the first nuclear submarine in 1952:

> The *Nautilus* will be able to move underwater at a speed of more than twenty knots. A few pounds of uranium will give her ample fuel to travel thousands of miles at top speed. She will be able to stay underwater indefinitely. Her atomic engine will permit her to be completely free of the earth's atmosphere.

The *Nautilus* went to sea on January 17, 1955. She soon broke every submarine endurance record for sustained high speed and distance traveled underwater. Because the *Nautilus* did not have to surface or raise a snorkel to obtain air for engine combustion, and because the crew's oxygen could be extracted from seawater, the *Nautilus* and succeeding atomic subs have been able to transit under the polar ice pack to the top of the world.

Today, submarines are a key component of all major navies. The U.S. Navy operates two principal types of submarines:

Crewmen load a torpedo into the nuclear-powered submarine *Nautilus*. *Attack* submarines carry torpedoes for use against enemy submarines and surface ships; the ballistic or *strategic missile* submarines carry the long-range Polaris/Poseidon weapons.

strategic missile submarines and attack submarines. (Figure 4 lists the overall missions of the Navy and the specific roles assigned to submarines.)

**Figure 4. NAVY MISSIONS AND SUBMARINES**

| Mission | Forces |
| --- | --- |
| Nuclear Deterrence | — POLARIS/POSEIDON MISSILE SUBMARINES<br>— SUBMARINE SUPPORT SHIPS OR TENDERS<br>— aircraft carriers |
| Sea Control | — surface warships, such as cruisers, destroyers, frigates, escort ships<br>— TORPEDO-CARRYING SUBMARINES<br>— small mine, patrol, and missile ships and boats<br>— aircraft carriers and their planes<br>— land-based patrol planes |
| Projection | — aircraft carriers and their planes<br>— amphibious ships, such as LSTs<br>— Marine landing forces<br>— surface warships |
| Presence | — most naval forces |
| Support | — auxiliary ships, such as tenders, repair ships, oilers, ammunition ships, tugs, etc.<br>— land bases, depots, shipyards<br>— research and experimental facilities and ships<br>— communications, intelligence, and weather forecasting activities, etc.<br>— schools and training centers |

The strategic missile submarines, armed with long-range Polaris and Poseidon missiles, are the nation's most-survivable deterrent against nuclear war. There are 41 of these underwater giants in service, each between 382 and 425 feet long, some 33 feet in diameter, and carrying a crew of about 120 officers and enlisted

The ballistic missile submarine *Henry Clay* fires a Polaris missile while on the surface. Normally these submarines are fully submerged when they fire missiles. Most of the submarine is underwater even when the 425-foot craft is on the surface.

men. Each submarine has 16 tubes for long-range missiles which can be fired while the submarines remain fully submerged. The design of the submarines has permitted periodic installation of more-advanced missiles as they become available.

As of mid-1974 the U.S. strategic submarine force consisted of:

Five submarines carrying sixteen Polaris A-2 missiles, each with a range of 1750 miles and armed with a single nuclear warhead.

Twelve submarines carrying sixteen Polaris A-3 missiles, each with a range of 2880 miles and armed with three small nuclear warheads which are "shot-gunned" down on a single target.

Twenty-four submarines carrying sixteen Poseidon C-3 missiles, each with a range of about 2880 miles and armed with ten small nuclear warheads that can be aimed at separate targets.

Each Polaris/Poseidon missile submarine carries sixteen long-range missiles. Here the *Sam Rayburn*'s missile tubes are open. In addition, the submarine is armed with torpedoes for self-defense. *Newport News Shipbuilding & Dry Dock Co.*

Nuclear-powered submarines, like most Navy ships, are fantastically complex. Here technicians aboard a Polaris submarine operate a missile launch console.

Eventually, all but ten of the forty-one missile submarines will have Poseidon (the other submarines are unable to accommodate a missile larger than the Polaris A-3). The Poseidon warheads are MIRV or multiple independently targetable re-entry vehicles. This means that after the missile boosters fall away, the front end, or missile "bus," spits out ten small nuclear "bombs" at separate targets. MIRV increases the probability of destruction of enemy cities—even if the enemy develops an antiballistic missile (ABM) defense.

It is significant that the Polaris and Poseidon weapons are for use against so-called "soft" targets, such as cities; they do not have the accuracy and other capabilities for use against enemy missiles, as would be required in a "first strike" weapon system.

The nuclear-propelled attack submarine *Ray* shown under way at high speed on the surface. Designed to operate underwater, these submarines seldom run on the surface. Indeed, they are faster traveling submerged than surfaced.
*Newport News Shipbuilding & Dry Dock Co.*

The attack submarines of the U.S. Navy are designed to seek out and attack enemy surface ships and submarines. This has been the traditional role of submarines since the century began. Their main weapons are "homing" torpedoes which are fired from tubes in the submarine. These self-propelled underwater torpedoes are guided to the target either by the sounds of the enemy's propellers or by sending out an acoustic pulse or beam that bounces back to a small receiver in the torpedo.

Modern torpedoes, like the Navy's Mark 48, can travel very

fast and can go thousands of yards. The Navy's newer submarines also can fire a guided missile called SUBROC (for SUBmarine ROCket). SUBROC is fired from a torpedo tube but streaks to the surface and flies through the air for several miles heading toward the location of an enemy submarine. Then the missile booster falls away and the payload—either a homing torpedo or nuclear depth charge—enters the water on a small parachute. Once back in the water, the parachute falls away and the torpedo seeks out the sub.

The World War II deck guns are no longer carried by U.S. submarines because of the vulnerability of a surfaced submarine to detection and attack, and the loss of speed caused by the under-water drag of the deck gun.

Modern U.S. Navy attack submarines are also noted for their advanced electronic detection equipment, or sonar, which can detect other undersea craft or surface ships at great distances. The ability of an attack submarine to operate silently in the same medium as the enemy sub makes it a particularly useful antisub-marine weapon.

Today the U.S. Navy has about fifteen diesel attack sub-marines, most of which will be phased out of the fleet in the near future. Some sixty nuclear-propelled attack submarines are in ser-vice, each manned by up to 110 officers and enlisted men.

Another twelve nuclear attack submarines are now under construction, and the Navy will operate a force of about seventy-five of these boats for the foreseeable future.

The modern nuclear submarine no longer resembles the "pig boats" described earlier by Harold Busch. Modern submarines are clean, air-conditioned, and comparatively roomy; individual bunks with private lighting and lockers are provided for each man. Some of the best food in the Navy is served to the crew in a combination dining-recreation area. Between meals, this area is used for exer-cise with special gymnasium equipment, movies, and the like.

Along with improvements in habitability, new submarines carry the latest in electronics for precise navigation, weapons control, communication, and detection. Sophisticated electronics

Nuclear-powered submarines are comparatively large and comfortable compared to earlier undersea craft, which were often referred to as "pig boats." This is the crew's dining space in the *Nautilus*.

serve as the primary "eyes" and "ears" of the submarine, making the undersea boat far safer for its crew and more dangerous for its enemies than ever before.

Submariners (pronounced "SUB"-marine-ers") are proud of the dolphin insignia on their chests, their special training and teamwork, their modern "boats," and a tradition for courage and skill. U.S. submarines compiled an impressive record during World War II and they continue this performance today. Whether they serve in diesel boats, fast, new, nuclear-powered attack submarines, or in the powerful Polaris/Poseidon submarines, today's submariners know that they are now and will continue to be important players on the nation's defense team.

Nuclear-powered submarines turn up in odd places. This is the sail structure of the *Whale* after the A-sub surfaced at the North Pole during an under-ice operation.

A Navyman stands by an antiaircraft missile during a check-out aboard a cruiser. He and his shipmates have undergone considerable training in modern weapons, electronics, hydraulics, and other specialized fields.

# The 9 Navy Today: People

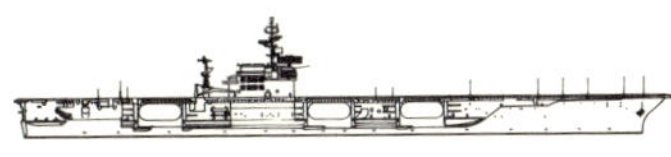



—ADMIRAL E. R. ZUMWALT,
U.S. NAVY, CHIEF OF
NAVAL OPERATIONS

The warship steaming out of the harbor looks impressive in the early morning light. She cuts the water with her bow, creating an ever-widening wake behind her. With her signal flags flying and her radar antennas rotating, the warship is a mixture of the old and new; she combines 200 years of U.S. Navy tradition and know-how with some of the world's most modern equipment.

But unless we are careful, we are likely to overlook the most important item on the warship—her people. Nuclear power,

Pride in the Navy: Two sailors polish the gold star awarded to their ship, the converted yacht *Venetia*, for her efforts against German U-boats in World War I. These men and their modern counterparts have helped this nation use the sea in peace and war.

complex electronic equipment, automatic guns, and guided missiles cannot replace men at sea. In fact, the introduction of some sophisticated hardware into the fleet has required even more human attention than before.

Throughout naval history, men have made the difference. They have snatched victory from the jaws of defeat; made important advances in engineering, armament, navigation, and exploration; and created the world's most powerful fleet. Navymen have made outstanding contributions to science and technology, and they have thus influenced man's future on this planet.

Men and women also make the Navy "go." Without their talents and dedication, ships would be nothing more than lifeless hulks of floating steel, airplanes would be reduced to expensive

Women are now serving in most Navy specialties. These Navy storekeepers are checking aviation parts listings; other Navy women serve aboard ship.

toys, and America's security would stop at the water's edge. But because of the knowledge and labor of the Navy's officers and enlisted men and women, the fleet sails in all of the oceans, its aircraft take off from carrier decks and airfields all over the world, and the Navy is able to perform its primary mission of safeguarding peace.

## ENLISTED MEN AND WOMEN

From the beginning, enlisted personnel have been the backbone and the muscle of the Navy. They perform an astounding variety of jobs, ranging from seamanship to operating computers. In fact, there are more than seventy different Navy job specialties for the enlisted man or woman (these are listed in Appendix B). But long before the sailor or "white hat" performs as a specialist, he must be trained. For this reason, the Navy runs 450 schools offering several thousand courses for more than three-quarters of a million students each year. A significant part of a sailor's life is devoted to schooling and training, and this education begins at a recruit training center, or "boot camp."

Proudly wearing their uniforms for the first time, the new sailors enter a seven-week period of rigorous training in many of the Navy's basic skills, such as seamanship, first aid, fire fighting and control of shipboard damage, swimming, pistol and rifle shooting, and marching. While in recruit training, enlisted personnel are given extensive tests to determine what Navy jobs they can do best. The results of these tests are matched with the sailor's own choices to help determine the specialty to which he is assigned. Boot camp is a challenge to the recruit because he has much to learn in a very short time. One out of ten fails to make it to graduation day, but those who do graduate receive an intense feeling of pride and accomplishment after it's over.

After graduation, the white hat receives orders to his first duty station. He might go to one of the Navy's many surface ships in home or foreign waters or he might be sent to further schooling. This training will be the beginning of an occupational specialty; the campus for this training is the Class "A" school. More than seventy Navy Class "A" schools teach the practical information and skills needed to advance to a particular "rating" or naval occupation. For example, there is an eight-week school for cooks and bakers (called "commissarymen") and a six-month course at the basic nuclear power school for reactor operators. This course, the longest for sailors, is followed by another six months of training using both classrooms and actual prototype nuclear reactors. Still further training may be necessary if the student wishes to serve aboard nuclear submarines.

Whether or not a man or woman (WAVE) attends Class "A" school, the Navy will be sure to provide on-the-job training at the first duty station. Under the watchful eyes of experienced petty

The Navy operates on, over, and under the sea. Here Navy divers check out an underwater installation. The U.S. Navy has been a leading developer of diving techniques and equipment.

officers, the green seaman is gradually taught all of the essentials of his new job. Soon he is given the opportunity to take a written and practical examination, and he begins to advance toward higher pay and rank.

Since many naval personnel wish to further their nonmilitary and academic educations while on active duty, the Navy provides

Big aircraft carriers and their jet-propelled aircraft are operated by men. Here a flight deck director guides an EKA-3B Skywarrior on the deck of a carrier.

Navymen live relatively well aboard ship and ashore. This is
the crew's mess compartment aboard the nuclear-propelled
missile frigate *Bainbridge*.

correspondence courses in a number of subject areas. These
courses are on the pre-high-school, high-school, and college
levels, and many sailors and WAVEs have received high-school
equivalency diplomas and college credits. The Navy also offers
educational benefits through the Off-duty College, Instructor Hire, Afloat College Education, Associate Degree
Completion, and Navy Enlisted Scientific Education programs.

The sailor's life is being improved in other ways, too. The
Navy has made reforms in shipboard living conditions, by giving
each man more space and privacy, lounge areas, and more at-
tractive messing and berthing facilities. This emphasis on habitability means that a ship can be both a home at sea as well as a
man-of-war. Navymen ashore are moving into modern housing

Responsibility rides heavy on the shoulders of naval officers. Here the commanding officer of a Polaris missile submarine directs his crew as they bring the "boat" alongside a submarine support ship. This officer is responsible for the 425-foot submarine, a crew of about 150 officers and enlisted men, and the nuclear weapons the sub carries. The responsibilities of command at sea demand men of outstanding capabilities.

where the emphasis is on good looks, privacy, and comfort. Both in the fleet and on Navy bases around the world, enlisted men are beginning to wear new stylish uniforms with up-to-date personal grooming.

Looking ahead to his future, the enlisted man sees that there are practical benefits to serving in the Navy. Some enlisted personnel are assigned to bases overseas, and they gain the chance to observe, participate in, and learn from life in another country. Also, much of the practical knowledge he acquires during his naval service is useful—even valuable—to civilian employers. While "outsiders" pay considerable sums of money to private schools to learn trades and professional specialties, Navymen are educated without charge. Some of the Navy's training, available to both men and women, is in the following fields: computer technology, oceanography, food service, electronics, photography, aircraft maintenance, journalism, medicine, and office or personnel administration. The list is almost endless and so are the rewards.

## THE OFFICERS

About one out of every ten Navy men and women is an officer. They serve in command or supervisory positions aboard ship, in aircraft, and on naval stations and shore bases. Some officers specialize in a number of shiphandling and warfare areas; they are designated "unrestricted line." Other officers—restricted line, staff, or corps officers—are experts in special naval activities, such as intelligence, law, supply, or engineering.

There are several paths to an officer's commission; some lead directly from civilian life, and others are designed to promote outstanding enlisted men into the officer ranks. The Naval Academy at Annapolis, Maryland, is probably the best-known route for both qualified high-school graduates and for sailors selected by competition within the fleet. After four years of demanding education, the Annapolis graduate receives a B.S. degree and a commission as an ensign in the Navy. The Navy also

The Navy's primary source of career officers is the Naval Academy at Annapolis, Maryland, which graduates almost one thousand ensigns each year. The young men complete the four-year program with a college degree as well as a commission in the Navy.

The Naval Academy—as well as many other Navy schools—provides some of the most advanced technical training available in the country.

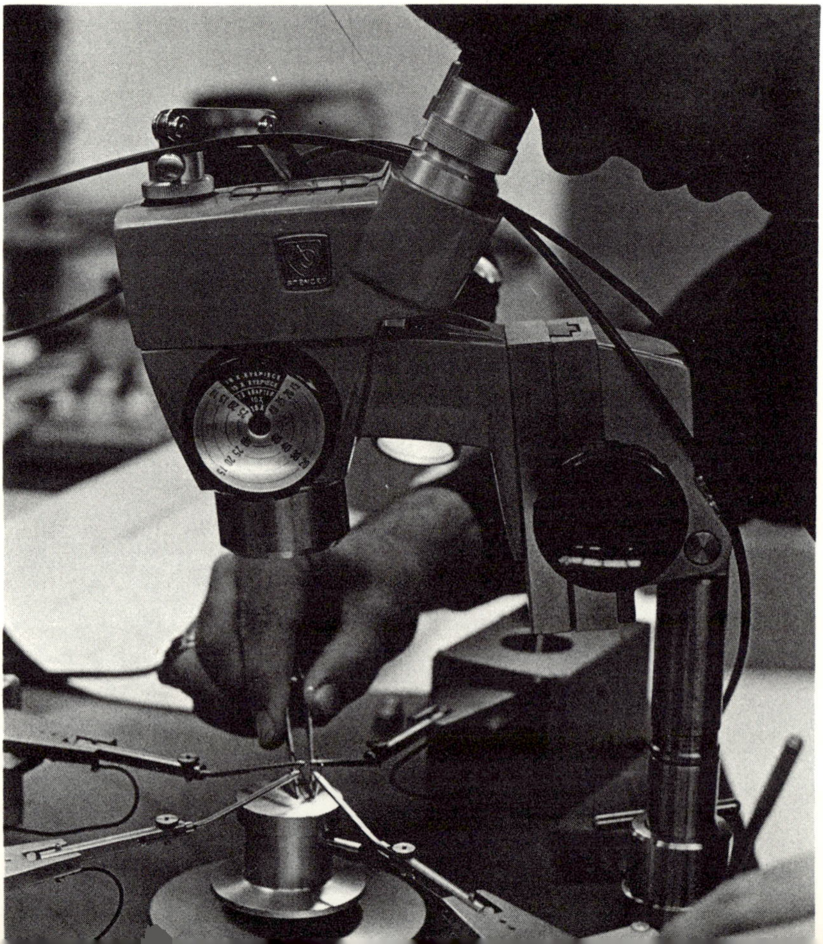

Naval aviators are considered the best fliers in the world . . . they have to be to land a jet aboard aircraft carriers at more than one hundred miles per hour, often at night, and while the flight deck is pitching and rolling. Here, in clear weather, a Navy A-4 Skyhawk is about to land aboard the carrier *Enterprise* while two nuclear-propelled missile ships steam behind the carrier.

has an Officers Candidate School (OCS), at Newport, Rhode Island, to train college men and women—either straight from the campus or from Naval Reserve units—in a sixteen-week course leading to a reserve commission.

Another program available to college students is the Naval Reserve Officers Training Corps (NROTC), located at about fifty campuses across the nation. NROTC enables a student to attend school and receive certain pay and allowances and officer training at the same time, with the Navy paying tuition and some expenses. Other options available for enlisted men include direct commissions as limited duty officers, or attendance at a four-year civilian university (followed by OCS)—all at Navy expense. These various programs, and others, ensure that the Navy's officer cadre receives young blood from different backgrounds. This keeps the officer corps fresh, spirited, and representative of the nation at large.

The new ensign's orders will send him to his first duty station or on to further schooling. Often the Navy will want to train an officer in a specialty, such as communications or weapons, before he reports aboard ship. After arriving at his initial assignment, the "boot" ensign will usually become a division officer, with a specific job such as assistant communications officer or damage control assistant, with responsibilities for the administration of a division of enlisted men with some twenty to several hundred personnel.

Some officers concentrate on one professional field and become experts in it. These men are the Navy's aviators, supply officers, doctors, lawyers, engineering duty officers, and submariners. Each contributes his talents, aptitudes, and special knowledge in a restricted area, but each is a very valuable member of the overall Navy team.

All officers aspire to "command," and work toward the day when they will "skipper" a ship, squadron, or shore facility. The road to command is a long and tough one because the Navy is highly selective when picking its captains. From his early days as an ensign, the officer must learn the science and the art of con-

Ships are run by muscles as well as brains. Here line handlers
aboard a destroyer pull in ammunition being transferred from
a replenishment ship steaming alongside.

trolling a warship at sea and in combat. He begins by standing
watches on the bridge, in the combat information center, and in
the engine room. Soon he learns how to maneuver his ship under
various operating and weather conditions. Later, if he performs
well, he is promoted and takes on wider duties, such as department head, and has other junior officers assisting him.

Promising officers are sent back to school later in their
careers to broaden their knowledge. Many attend the Navy's
Postgraduate School at Monterey, California, where they may
receive masters' degrees. More senior officers might study at the
Naval War College in Newport, Rhode Island, to prepare for
high positions within the Navy or in other defense and government agencies. In addition, every year many naval officers attend
leading civilian universities to obtain advanced degrees in fields
of interest to the Navy.

Regardless of whether they are enlisted or commissioned,
Navy men and women move through a cycle of tours ashore and

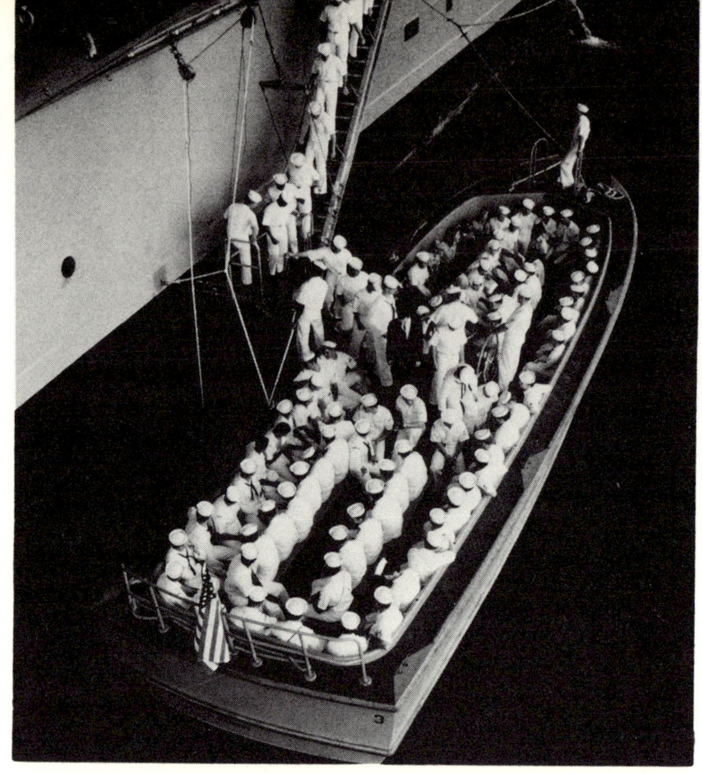

Navymen go to sea and ashore. These sailors from the carrier *John F. Kennedy* are going ashore for liberty in Montego Bay, Jamaica.

afloat. They serve in shipyards and bases in the United States, aboard destroyers in the Pacific and on giant carriers in the Mediterranean, in patrol aircraft over the Atlantic, and aboard missile submarines sailing from Spain or Scotland.

Navy people are men *and* women. They are white, black, brown, and yellow. They come from the nation's inner cities, suburbs, towns, villages, hamlets, and hollows. Crews of officers, warrant officers (career specialists), chief petty officers, petty officers, and enlisted men and women man the complex surface ships, aircraft, submarines, and shore stations of the Navy. On a daily basis they often have more adventure and interesting work than many people have in a lifetime.

At this moment, Navymen are handling small boats, jet planes, computers, stethoscopes, bulldozers, cruisers, cameras, paintbrushes, radios, and missiles. Today's Navyman and woman

Ship, men, and the sea: America has traditionally been a seafaring nation. Today's Navymen carry on the traditions of John Paul Jones, Dewey, Peary, Byrd, Halsey, and Nimitz.

are trained in subjects beyond their grandfathers' wildest dreams. All of this knowledge, equipment, and capability have been made by and for some of the world's most talented men and women—the people of the U.S. Navy.

The U.S. Marine Corps has demonstrated in war, as on Iwo Jima where this historic flag raising took place, as well as in peace, that "soldiers of the sea" are invaluable to a maritime nation.
                                                        *U.S. Marine Corps*

# 10 The Leathernecks

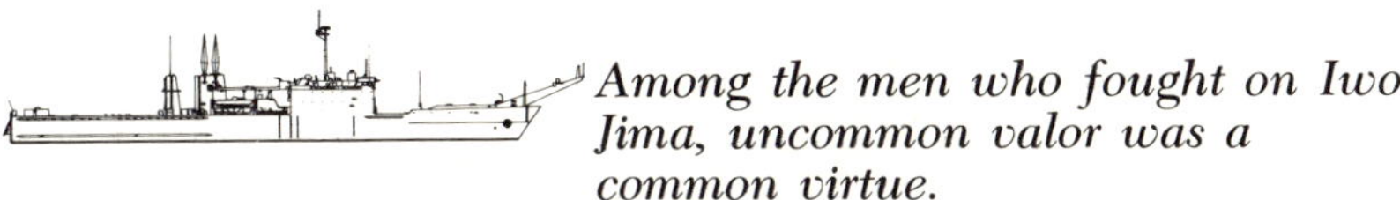

*Among the men who fought on Iwo Jima, uncommon valor was a common virtue.*

—FLEET ADMIRAL CHESTER W. NIMITZ, U.S. NAVY(1885–1966)

They are as at home on the deck of a warship as they are on jungle patrol. They belong to the Department of the Navy but remain separate from the Navy. They fly supersonic aircraft, drive tanks, and fire missiles, but they also perform long, forced marches. They are Marines.

The Marines are many things. Primarily, they provide the U.S. Navy with an amphibious assault capability for landing on enemy beaches. In providing this capability, the Marines are among the most versatile, hard-hitting, and élite military forces in the world.

# HISTORY

The Continental Marines were established in November 1775. During the American Revolution these Marines fought from the masts of American warships, firing rifles and dropping grenades onto the decks of British frigates. Their volleys killed many English gunners and officers. Marines also got their first amphibious landing experience during the Revolution. In February 1776, for example, two hundred Marines were landed at Forts Montague and Nassau in the Bahamas to capture badly needed gunpowder for General Washington. After Britain's defeat, the Marines passed out of existence. The Corps was reestablished in July 1798, and has remained part of the U.S. armed services ever since.

Marine officers and men have participated in every war fought by the United States, including the Quasi-War with France, the Barbary War, the War of 1812, various Indian campaigns, the two-year conflict with Mexico in the 1840s, the Civil War (both the North and South used Marines), the Spanish-American War, the insurgencies in the Philippines (1898–1902) and China (1900), both world wars, Korea, and Vietnam.

The World War II fighting in the Pacific saw some of history's toughest battles, with the Marines in the thick of it. New amphibious warfare tactics, developed by the Marines, called for the troops to be carried out to the invasion area in large Navy transports. They would then disembark into smaller landing craft, and head toward the island in "waves." At the same time, other Navy warships were to bombard the target with gunfire and rockets, to kill defenders and smash their fortifications. Then other specialized assault vessels—such as the famed LST (for landing ship, tank, or sometimes, affectionately, "large slow target")—would move up to discharge tanks, artillery, and other combat gear directly onto the beach. Often, the Marines fought hand to hand with the Japanese defenders to drive them from their dug-in positions in caves, pillboxes, and bunkers. This kind of difficult warfare is symbolized by one of the proudest moments in the Corps' history—the raising of the Stars and Stripes over Mount Suribachi on Iwo Jima in 1945.

In more recent times, the Marine Corps served in Vietnam in the nation's longest war. While there were no large-scale landings as in World War II, the Marines were given major responsibilities for defending Vietnam from Communist insurgency. They were deployed in the northern area of the country, and fought the Vietcong and North Vietnamese at such places as Khe Sanh, Hue, Quang Tri, and Con Thien. As in times past, Marines were among the first on the scene. Marine helicopters and antiaircraft missile units were among the first U.S. units sent

The U.S. Marine Corps has long experience in "limited war" conflicts in Asia and Central America. This camouflaged Marine was an adviser to South Vietnamese forces during the Vietnam War. Under U.S. Marine guidance, South Vietnam has developed a most effective marine force.

to Vietnam to support and defend U.S. advisers to South Vietnamese forces. They later fought as ground troops and also conducted several small-scale amphibious landings.

# ORGANIZATION

The Marine Corps occupies a unique position within the Department of Defense. The head of the Corps, the commandant, is a general, appointed by the President, who reports to the Secretary of the Navy regarding the administration, training readiness, and organization of the Marines. The Navy's Chief of Naval Operations controls Marine units which are assigned to major fleet commands, as well as to individual warships. The commandant meets as one of the Joint Chiefs of Staff whenever the Army, Navy, and Air Force are discussing matters directly concerning the Marine Corps.

Internally, the Marine Corps is made up of Fleet Marine Forces (FMF), forces afloat, and security forces. The Fleet Marine Forces are the heart of the Corps; there is an FMF assigned to the Atlantic Fleet and to the Pacific Fleet. The FMF

The two Fleet Marine Forces, FMF Atlantic and FMF Pacific, are balanced forces with infantry, artillery, and tank components. Here Marines drive an M103 heavy tank aboard a Navy LST.

Tools of the trade for Marines include amphibious tractors or "amtracs" which gained public attention during the classic assaults against coral atolls in the Pacific during World War II. Here, tractors, officially designated LVT (landing vehicle, tracked), are being driven aboard a Navy landing ship during exercises.

consists of Marine companies, battalions, regiments, divisions, aircraft wings, and support forces. As required, different Marine units are organized into teams, such as an 1800-man battalion landing team which has its own tank, artillery, amphibious tractor (amtrac), and supply components.

The Marine air wings are unique; the U.S. Marine Corps is the only such force in the world with its own air arm. Each wing has fighter, attack, reconnaissance, transport, and helicopter squadrons, with all but the transport aircraft being able to fly from Navy aircraft carriers and airfields ashore.

As pioneers in the use of helicopters and dive bombing to support Marines in land warfare, the Corps is now flying the vertical/short take-off and landing (VSTOL) aircraft. Current

The U.S. Marine Corps is believed to be the only marine force in the world with its own air arm. Marine pilots, who are *naval* aviators, fly modern fighter and attack aircraft, and a large number of helicopters, such as this CH-46 Sea Knight about to land on a Navy amphibious ship. *Boeing Company*

planning provides for one squadron of AV-8 Harrier VSTOL strike aircraft in each of the three aircraft wings.

## MISSIONS

Any organization that is prepared to "fight in the air, on land and sea," as is the Marine Corps, is bound to have many missions. In general, the Marines serve as a ready, quick-reaction force for use in emergency or limited war situations. They will almost always come by sea, using helicopters, landing boats, and amphibious tractors to get ashore.

Waiting for the signal to come ashore, Marine-carrying landing craft circle in holding patterns during an amphibious
exercise. Marines also come ashore in amphibious tractors and
helicopters during assaults.

The World War II method of sending troops to the beach in
a "horizontal" assault with small boats has been supplemented by
"vertical" assault. This "up, over, and down" tactic uses troop-
carrying helicopters to lift the Marines from ships to landing
areas behind the enemy's coastal defense positions. This allows a
two-pronged attack: a frontal thrust comes on the surface while
a second drive comes from the rear after arriving by air. Still,
Marines must always capture and hold the beaches because almost all of the tanks, supplies, ammunition, and support personnel will arrive across the beach.

Marines are assisted in amphibious assaults by prelanding

shore bombardment from cruisers and destroyer-type ships. Carrier-based aircraft perform close air support missions. After the Marine assault units have established a secure perimeter, "portable" airfields and Marine air units are set up ashore. Marine fighters, attack planes, and helicopter gunships hit the enemy's supply and communications lines to destroy his capacity to fight, and give close fire support to Marine ground units. In addition, Marine aircraft ferry in their own supplies of ammunition, food, and equipment, and can either off-load the material at smooth or rough airstrips, or parachute it to troops in the field.

Other Marine missions are internal security and local defense for naval bases and stations at home and overseas. Marines guard special government facilities and installations, especially embassies, consulates, and legations, on foreign soil. "Leathernecks" (the name comes from the stiff collar on old Marine uniforms) in the forces afloat serve as gun crews and landing parties aboard aircraft carriers and cruisers.

## THE FUTURE

Essentially, the Marine Corps is a specialized sea-based army which can land ashore in a short time. There has been a growing reliance on U.S. forces afloat because American forces are being withdrawn from fixed bases in foreign countries.

The Marines require little outside support, making them ideal for crises. Because they are experts at living and working under rugged conditions, the Marines do not need elaborate, fixed bases; they stay in an area just long enough to accomplish their mission and then they are gone, as in the Lebanon and Dominican Republic landings. Because the Marine Corps is a relatively small military force, every man in it must be in excellent physical shape and very proficient in combat. Many of the people who "support" Marines, such as doctors, dentists, corpsmen, and chaplains, are Navymen. Thus *every* Marine can be a

On the beach, Marines rush inland to secure a beachhead.
Marines are equally at home at sea or ashore, and serve with
the Navy aboard ship and at land bases.

fighter; this attitude is reflected in the slogan, "the Marines need
a few *good* men."

Now, almost two hundred years after their founding, the
history, traditions, accomplishments, and—most important—
capabilities of the Marines make them key members of today's
Navy.

Ocean research is a tedious and often dirty job; but knowledge of the oceans is important for commercial and military use of the sea. Here a sailor aboard a Navy surveying ship prepares a device for measuring water temperature.

# 11 Science and the Navy

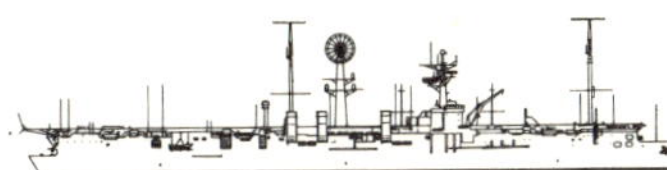

*Knowledge of the oceans is more than a matter of curiosity . . . our very survival may depend on it.*

—JOHN F. KENNEDY (1917–1963)

How a nation uses the seas around it determines in many ways that country's security, wealth, international relations, and quality of life. The use of the sea—the broad oceans covering almost three-quarters of the earth—depends upon man's knowledge in several fields, including oceanography, navigation, marine engineering, and communications.

The U.S. Navy seeks to learn all that it can about the seas, and has long been a major contributor to ocean research and related forms of science. Navy laboratories, research offices, oceanographic ships, and surveying vessels are dedicated to discovering, gathering, studying, and applying new information. Some of this knowledge is used in developing military equipment, but much of it is also valuable to civilian scientists and commercial industries.

Most of the Navy's scientific work involves long hours of tedious, patient research and study. There are few dramatic moments, and the men and women who perform this work are not heroes in the popular sense. This is regrettable, since everything that man knows is the combination of all of the scientific "production" in history. Great breakthroughs are only possible because the track was marked by earlier scientists and engineers. Still, the Navy has had its scientific "heroes" and its own share of important breakthroughs.

## EARLY PIONEERS

One of the earliest heroes was Lieutenant Charles Wilkes, who, in 1838, led the first scientific expedition ever sponsored by the American government. Wilkes took six ships down to the Antarctic continent, exploring more than 1500 miles of its coastline before returning to the United States in 1842.

In another field of science—physics and marine engineering—the Navy was moving into new areas. The first naval steam-powered ship was the U.S.S. *Demologos,* launched in 1815. The second Navy vessel to be powered by steam was the *Fulton II,* built in 1832. Her construction was supervised by Commodore Matthew Calbraith Perry, who is famous for opening Japan to Western commerce. Perry took command of the *Fulton II* and used her to train Navymen in steam engineering. He spoke to Congressmen on behalf of oceangoing steam ships and was a major contributor to the "Steam Navy."

One of the greatest advances in man's knowledge of the seas was made by Lieutenant Matthew F. Maury, who entered the Navy in 1825. He soon became an expert in navigation, gathering information on weather, water, and winds from every possible source, including daily observations sent to him by ship captains and navigators, and he put together wind and current charts. He established the world-famous Navy Hydrographic Office and the

Navymen have been in the leading ranks of world explorers since the United States was established. Robert E. Peary, shown here during a polar expedition, was one of the first men to lead explorers to the North Pole. Another naval officer, Richard E. Byrd, was the first man to fly over both poles and also spent an "Antarctic night" *alone* at the South Pole.

Naval Observatory. This Naval officer-scientist finished his career in the service of the Confederate States Navy after deciding to return to his native Virginia during the Civil War.

Perhaps the Navy's most famous scientist was Albert A. Michelson. Formerly a naval midshipman (he entered the Naval Academy at sixteen), he became an instructor of physics and chemistry at the Naval Academy in 1875. His studies included the tides and astronomy, but his most famous accomplishment was the accurate measurement of the speed of light. He won history's first Nobel Prize for physics in 1907.

Another Navy pioneer was Lieutenant (later Rear Admiral) Robert E. Peary who determined that Greenland was the world's largest island and not, as had been believed at the time, part of a continent. Peary's most famous accomplishment came after a rugged fifty-day trek across the frozen northern ice when he reached the North Pole on April 5, 1909.

# TWENTIETH-CENTURY SCIENCE

After the turn of the century, aviation began affecting both warfare and science. One of the first demonstrations of air flight's global implications occurred in May 1919 when Navy Lieutenant Commander Abner C. Read and a crew of five flew the seaplane NC-4 on history's first transatlantic crossing by air. The flight, between Long Island, New York, and Lisbon, Portugal, pushed the development of better aircraft, radios, charts, and other equipment.

Commander Richard E. Byrd found that the airplane was a good vehicle for exploration. On May 9, 1926, he flew over the North Pole, and symbolically dropped a medal worn by his predecessor, Peary. He then turned his attention southward, and directed the first airplane expedition into Antarctica. As a result of his pioneering flights, Byrd, later promoted to rear admiral, was for many years the only man ever to have flown over both poles.

The efforts of two men—Dr. Ross Gunn and Captain (now Admiral) Hyman G. Rickover—moved the Navy into the atomic age. Dr. Gunn, who was a research supervisor at the Naval Research Laboratory, recognized the potential of nuclear energy. Starting with a modest $1500, Gunn began research into the use of nuclear energy for submarines in 1939! This was two months before Albert Einstein's famous letter to President Roosevelt proposing the atomic bomb.

Rickover, a brilliant manager, directed the development of a practical submarine nuclear power plant. Thus, the U.S.S. *Nautilus* got under way on January 17, 1955, in man's initial use of atomic energy to produce propulsion.

In the mid-1950s, Commander George F. Bond, a Navy physician and submariner, performed research aimed at combating the "bends," an ailment affecting divers. The bends is triggered when divers come to the surface too rapidly. Before Dr. Bond's work, divers had to be raised to the surface slowly, even

The U.S. Navy began the development of nuclear propulsion for ships in the late 1930s, even before the atomic bomb program was initiated. After World War II the U.S. Navy developed nuclear propulsion for submarines and surface ships under the astute leadership of H. G. Rickover, shown here on the deck of the pioneer A-sub *Nautilus*.

Man descended to the deepest known point in the ocean— 35,800 feet—in the Navy bathyscaph *Trieste* when Lieutenant (now Commander) Don Walsh and Jacques Piccard rode the craft into the depths on January 23, 1960. Extensively rebuilt after the record dive, the "successor" *Triestes* have made valuable contributions to our knowledge of the oceans.

after only a few minutes of work at deep levels, to avoid the bends. Dr. Bond developed the "saturation" diving technique, in which helium instead of nitrogen is used in the diver's air supply. This permits divers to work for much longer periods on the ocean floor and reduces the risks.

Navy and civilian scientists have a great need to learn about the ocean bottom because so much of the "earth" is under water. The deepest section of the sea floor escaped exploration until January 23, 1960, when Lieutenant (now Commander) Don

This is an overhead view of the pressure sphere or gondola of the bathyscaph *Trieste*, showing the 5½-foot-diameter compartment that holds two or three crewmen plus equipment when the craft dives. In many ways the exploration of "inner space" has been similar to the exploration of "outer space," with the U.S. Navy prominent in both realms.

Walsh and civilian Jacques Piccard piloted the research submersible *Trieste* to a depth of 35,800 feet. That dive established a world record which still stands.

Navymen have also explored the outer reaches of the earth—and beyond. More than half of the nation's astronauts have been naval aviators. For example, Rear Admiral Alan Shepard was the first American in space, making a suborbital flight in May 1961; Marine Colonel John Glenn, now retired, became the first American to orbit the earth in February 1962; former Navy

Commander M. Scott Carpenter was the second American to orbit the earth in the Mercury project. Subsequently he became a Navy "aquanaut," testing advanced diving techniques in the SEALAB experiments. Here Carpenter poses with a model of a SEALAB sea-floor laboratory. In a similar structure, he lived twenty-eight days on the ocean floor at a depth of 205 feet.

Captain Walter Schirra flew in the Mercury, Gemini, and Apollo programs; and the man who took the "giant leap for mankind," Neil Armstrong, is a former Navy pilot.

In today's Navy, scientists, technicians, engineers, and sailors are hard at work in a multitude of scientific fields and with some of the most sophisticated equipment in the world. For example,

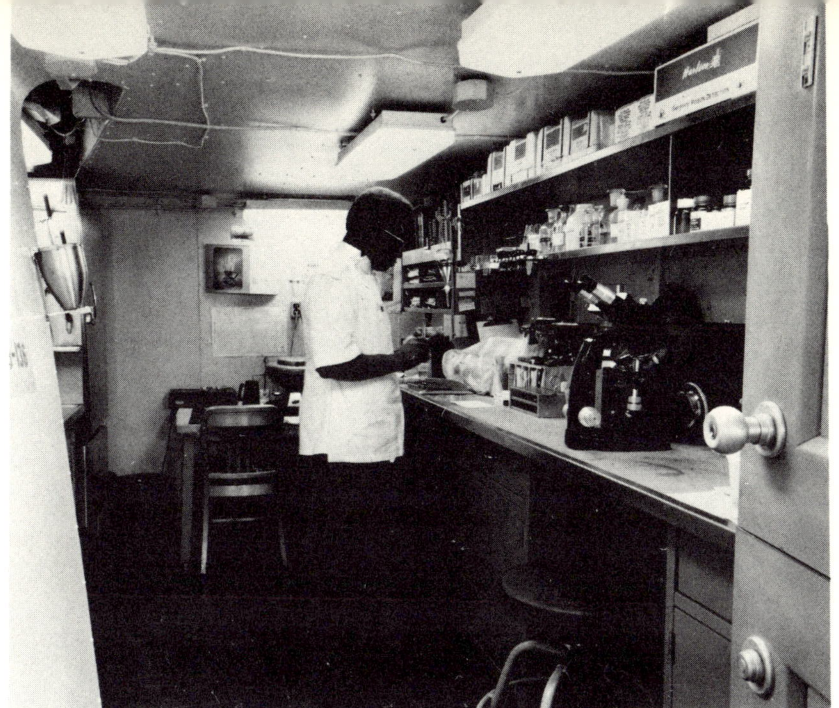

A Navyman in the medical laboratory of the aircraft carrier *John F. Kennedy* is working with some of the most advanced equipment and techniques in his profession. Other Navy medical personnel are today engaged in underwater and aerospace medicine, and the unique Navy Medical Research Units are developing cures for diseases in remote areas of the world.

Navy oceanographers use submersibles, surface ships, aircraft, and even satellites to study the oceans to provide data necessary for naval operations on, over, and in the oceans.

Similarly, Navy acoustic specialists are developing new electronic equipment and techniques for detecting submarines, communicating with ships thousands of miles away, and obtaining navigation data from satellites.

Medical personnel in the Navy are searching for new methods of combating disease in remote areas where naval forces may have to operate, better methods of treating combat casualties, and protecting pilots from the rigors of sustained supersonic flight.

The lists of fields of interest to the Navy are almost endless, for naval activity encompasses all media—air, land, and sea, and virtually all human endeavors. Man's increasing reliance on the sea for food and mineral resources, defense, and even recreation adds a new importance to the Navy's scientific programs. The realm of science, therefore, may well be the most essential field of Navy effort in the future.

# 12 The Competitors

*History shows that states which do not have naval forces at their disposal have not been able to hold the status of a great power for a long time.*

—ADMIRAL SERGEI G. GORSHKOV,
COMMANDER-IN-CHIEF OF
THE SOVIET NAVY

The United States is not the only nation seeking the advantages that accrue to those who use the sea. Other nations have recognized the advantages of becoming a sea power, and two of these, Japan and the Soviet Union, are vigorously pursuing maritime policies. Their increasing use of the sea for military, commercial, and political purposes means stiff competition for the United States in an age when man is probably more dependent upon the sea than ever before.

Japan and Russia have different reasons for wanting to become sea powers and each clearly has the will and the means to do so. It is significant that both countries began their maritime programs only recently. Japan's navy and merchant marine were totally destroyed during World War II, and Russia's naval and commercial fleets were almost as bad off. Yet today the two nations have built up to such a point that they are fierce political and economic competitors of the United States.

## JAPAN AT SEA

Almost thirty years after being defeated in one of the greatest naval wars in history, Japan has again become a sea power. For example, her shipbuilding industry, one of the keys to successful and continuing maritime programs, is the world's largest. Also, the Japanese merchant marine is one of the largest afloat, and is growing faster than most. And, finally, more fish are caught by Japan than by the United States.

The Japanese "Navy"—officially called the Maritime Self-Defense Force—is a small fleet compared to the U.S. and Soviet navies. Indeed, as is shown in Figure 5, it is lacking the larger warships seen in many fleets, such as missile submarines, aircraft carriers, cruisers, and amphibious ships. But the Maritime Self-Defense Force has been built in Japanese shipyards for the most part, and it is growing. The increase in size is matched by improvements in quality. The Japanese are building relatively large helicopter-carrying destroyers plus other destroyers armed with missiles.

The Japanese have a compelling interest in the sea because their industries are tied to overseas sources of oil and other vital supplies. For example, 99 percent of the petroleum used by Japan is brought to the island nation by ship. For decades they have had to engage in waterborne trade in order to survive as a modern people. Japan's success as a leading industrial nation and economic power has depended on the flow of goods and resources between its piers and the ports of other countries. This has required a strong merchant marine. It follows, therefore, that

One of Japan's supertankers at sea. This ship, the *Japan Carnation*, carries about ten times the petroleum of a World War II-era tanker. Petroleum carriers twice this size are in service, with most of the world's supertankers being built in Japanese shipyards. *Hitachi Suzen Shipbuilding*

Japan requires naval forces to protect its coasts and sea lanes from attack.

Not only do the shipyards of Japan turn out quality ships for her own navy and merchant marine, but many foreign companies, American firms included, buy tankers and other ships "made in Japan." Once the butt of jokes in this country, Japanese production is now recognized both here and abroad as being of superior quality.

As Japan moves into the mid-1970s, it will probably maintain the tempo of its maritime expansion. The Japanese Navy will continue to increase gradually, although the reduction of U.S. naval forces in the post-Vietnam era may produce a more vigorous Japanese approach to naval affairs.

### *Figure* 5. COMPARATIVE NAVAL FLEET STRENGTHS*

| Type | United States | Soviet Union | Japan |
|---|---|---|---|
| Strategic Missile Submarines (nuclear) | 41 | 40+ | — |
| Other Submarines (nuclear) | 60 | 70+ | — |
| Submarines (diesel-electric) | 12 | 200 | 11 |
| Aircraft Carriers | 14 | — | — |
| Helicopter Carriers | 7 | 2 | — |
| Cruisers and Frigates (missile-armed) | 38 | 40 | — |
| Cruisers (all-gun) | 1 | 10 | — |
| Destroyers (missile-armed) | 29 | 20 | 1 |
| Destroyers (all-gun) | 70 | 40 | 29 |
| Ocean Escort Ships | 65 | 110 | 14 |
| Amphibious Ships | 64 | 80+ | 4 |
| Patrol Ships and Craft | 13 | 500+ | 30 |
| Mine Warfare Ships and Craft | 33 | 280 | 50 |
| Underway Replenishment Ships | 61 | 50 | — |
| Support Ships | 130 | 150+ | 10 |

*Active ships. Numbers rounded for comparative purposes. U.S. Navy destroyer and mine sweeper totals include ships manned by mixed active-reserve crews and assigned to Naval Reserve training.

# RUSSIA AT SEA

The Soviet Union is the world's largest nation (almost three times the size of the United States), and only two countries—China and India—have more people. For many years, therefore, the Soviet Union has been regarded by world leaders as a *land* power. Never, during all of the centuries of Russian history, could the tsars or commissars realistically claim to be the masters of the sea—until recently.

After World War II, the Soviet leadership, first Joseph Stalin and then Nikita Khrushchev, began and maintained a dramatic naval buildup. Khrushchev's policies of the mid-1950s differed with Stalin's emphasis on a big surface navy by stressing the development of nuclear-propelled submarines and antiship missiles. In addition, the Soviets started on a massive development of the nation's merchant marine. After Khrushchev fell from power in 1964, his cosuccessors, Leonid Brezhnev and Aleksei Kosygin, supervised the introduction of ballistic missile submarines, similar to the U.S. Polaris.

A key man in this Soviet naval buildup has been Sergei Gorshkov, the commander in chief of the Soviet Navy and a deputy minister of defense since January 1956. Since then, when the relatively young Gorshkov became both the Soviet "Chief of Naval Operations" and "Secretary of the Navy," the U.S. Navy has had five admirals and nine secretaries in these two positions. Thus, Gorshkov has survived the political crises of Khrushchev's downfall in 1964 and at least two major military leadership "reorganizations," to build the first Russian fleet that can effectively challenge the West on the high seas.

Under Gorshkov's leadership, the Soviet Navy has developed the world's largest surface and submarine fleets, with more nuclear-propelled submarines than the U.S. Navy. Further, Soviet ships and submarines are of very high quality, and the Red Navy is the world leader in several fields, such as antiship missile development.

A Soviet Yankee-class submarine at sea. Since 1967 the Soviet Navy has built more than thirty of these nuclear-powered, strategic missile submarines. They have now been replaced on the building ways by the large Delta-class submarine which carries missiles that can travel more than 4,000 miles.

The Soviet Navy is the world leader in antiship missile development. This is an "Osa"-type missile boat firing the short-range (23-mile) Styx missile. Styx missiles from Soviet-built craft sank the Israeli destroyer *Elath* in 1967.

Similarly, the Soviet merchant marine is one of the world's largest and fastest growing; the Soviet fishing fleet—the world's largest—ranks third among fishing nations and pulls about three times as much seafood from the sea than is caught by the United States. The Soviet ocean research fleet is the world's largest, as is the Soviet fleet of space support and tracking ships. A modern and large shipbuilding industry backs up this broad thrust to the sea. For example, one of several Soviet submarine-building yards has a greater capacity for producing A-subs than all of the world's other shipyards combined.

All of these fleets are supported and supervised at the very highest levels of the Soviet government. For example, all Soviet ship construction, military and commercial, is under a single Ministry of Shipbuilding. Nothing like that overall coordination exists in the United States. In general, the Soviet government has

Today's Soviet warships are highly innovative, such as this antisubmarine helicopter carrier/missile cruiser. The 645-foot ship is a missile cruiser forward and a helicopter carrier aft; larger aircraft carriers will follow the two ships of this design.

made it a national policy to build powerful surface warships, submarines, merchant ships, and aircraft to make Russia into a modern sea power.

Now that the Kremlin has created a strong navy, it is boldly sending its warships into distant seas. No longer a coastal force, the new Soviet fleet flexes its muscles far from the Russian homeland. The Soviet flotilla in the Mediterranean Sea, today even outnumbers the U.S. Sixth Fleet from time to time. The warm waters of the Indian Ocean, Gulf of Guinea, and the Caribbean continuously feel the presence of Soviet naval units. Russian fishing trawlers—and intelligence ships—sail near the coasts of the United States, and the Soviet merchant marine calls at scores of ports in Africa, Latin America, Europe, and Asia

Soviet use of the sea includes a large and modern merchant fleet which has supported Soviet interests at great distances from the U.S.S.R., projecting Soviet power into areas such as Cuba and North Vietnam. This is the *Krasnograd,* one of the Soviet merchant ships loading grain at U.S. ports. By 1980, the Soviet Union will probably have the world's largest merchant fleet, except for supertankers. *Port of Houston*

This jet-propelled Badger of the Soviet naval air arm is overflying a U.S. aircraft carrier in the north Pacific. This plane is configured for reconnaissance; the Soviet Navy has some 285 of these Badgers that carry antiship missiles.

The Soviet Union has the world's largest fishing fleet and brings in more food from the sea than any nation except for Japan and Peru. Flotillas of small fishing trawlers operate for sustained periods in remote ocean areas with the support of large fish "factory" ships such as the *Baltisnaya Slava,* shown here off the coast of Virginia.

During the 1960s, the Soviet Navy began long-range operations on a scale unprecedented in Soviet history. This missile-armed cruiser and the submarine behind her were part of a Soviet task force that came within sight of the Hawaiian Islands during maneuvers in the Pacific Ocean in 1971.

(and, since 1972, even in the United States). Thus, in twenty-five years, the Soviet Union has become a land power which is equally at home on the ocean. The Russian bear has learned how to swim.

The United States is not the only country in the world with a dependence on the sea. Other nations, such as Spain and Holland, once carried on great international trade by using the ocean highways, while others, such as England and France, extended their influence through naval strength. In the twentieth century, which has been called the American century, the United States became a world leader, largely because of its ability to use the sea. Now, in the 1970s, the United States is being challenged on the ocean by Japan, the Soviet Union, and other countries. The future role of the United States in world affairs depends heavily on how well America and its navy respond to these new competitors.

The officers and enlisted men who operate the new ships of the
Soviet Navy are intelligent and highly motivated. These sailors
from a Kashin-class missile frigate are on liberty in a Middle
East port.

Most U.S. political and even economic programs in the 1970s and 1980s will be carried out under the umbrella of nuclear deterrence. This is an artist's view of the Trident missile submarines planned by the U.S. Navy which will be the nation's primary "weapon" for nuclear deterrence.

# 13 The Future

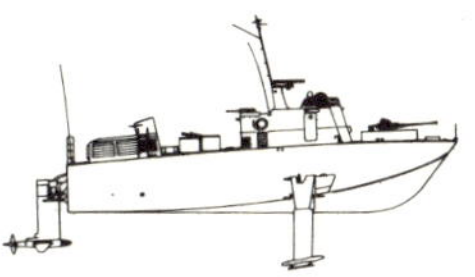



Man first went to the sea to fish and trade, and, in time, he sailed farther and farther from his home shores. The increase in distance meant man had to have bigger and stronger ships. It also meant that fighting ships were needed to protect the fishermen and traders from pirates and enemies. Over the centuries man has been changing his commercial and naval ships to keep up with the changing times.

Science and technology have given man the tools to improve steadily his ability to sail and fight upon the seas. Hulls, first made

145

out of bark, reed, or wood, were later built of iron and then steel, and are now also constructed of aluminum and fiber glass. Propulsion has been provided by oar and sail, then by burning wood, coal, and oil, and now nuclear energy. Combat at sea started with arrows, spears, and cutlasses, then moved into various forms of cannon and guns, and currently involves aircraft and missiles. The crow's-nest high atop the mast of a sailing vessel has been replaced by satellites and radars and computers.

Most of these changes affected the U.S. Navy; many of these changes were caused by the U.S. Navy. The Navy of the 1980s is under construction and the Navy of the twenty-first century is already on the drawing boards. The missions which will be assigned to the future Navy will determine, to a large extent, the kinds of ships it will have.

## TOMORROW'S MISSIONS

Today's world—its politics, economies, and geography—and the Navy's ability to sail in international waters relatively free of attack or restrictions make the ocean a vital area for the future. The Navy has performed well in a number of missions and political and military situations, and it will assume greater national importance as it meets future challenges. The ocean, by its vast area and because it washes the shores of the continents, is the stage on which national power will be shaped in the coming years.

This has already been recognized in the Nixon Doctrine, an American policy which calls on U.S. allies to take on a greater role in their own defense. As the allies increase their own military strength, the United States will be able to withdraw most of its air and ground forces from land bases in those nations. Thus, the U.S. Navy's strategic missile submarines, aircraft carriers, and amphibious ships will become increasingly valuable.

In addition, the United States and her allies in Europe and Asia are becoming more and more dependent on overseas supplies of oil and other important natural resources. By 1985 or so, the United States will have to import about half of the petroleum it needs, and this will require from several hundred to over a thousand tankers. Put another way, the U.S. requirement for oil in the future will be equal to the carrying capacity of all of the world's tankers afloat in 1970.

The sea can be a three-dimensional battleground in which nations send various missile-launching "platforms" above, on, and below the surface. Nuclear-armed missiles can be moved from place to place hundreds of feet beneath the waves. With fewer and fewer land installations at which to base its aircraft and armies, the United States will look to the sea as an avenue for projecting its influence and power across the globe. The events of the 1960s have made Americans anxious to avoid long and costly stays in foreign lands; so it must be remembered that sea-based forces are cheaper and safer to maintain than ground-based forces.

Tomorrow's naval missions will be similar to today's—deterrence, projection, sea control, and presence—but they will have greater importance than ever before, and they will be performed with new kinds of ships and by new kinds of sailors.

## TOMORROW'S NAVAL FORCES

The Navy's planners, strategists, and engineers are designing ships to perform the vital jobs of national defense in the late 1970s and beyond. Some of these ships are advanced and improved versions of current types, such as the nuclear-powered aircraft carrier, while others are revolutionary new types, such as the surface effect ship.

The Navy is building two nuclear-powered *Nimitz*-class carriers, with a third to follow. These ships will give the Navy twelve modern aircraft carriers of post-World War II construction. A new fighter aircraft, the F-14 Tomcat, is being produced to fly with the 100-plane air wings which will be carried by these powerful ships.

Two different types of submarines will join the fleet in

The sea control ship will be a small "aircraft carrier" operating helicopters and VSTOL aircraft in areas where there is comparatively "low threat" from enemy naval forces. Aircraft carriers would still be required when the enemy could bring major air, surface, or submarine forces to attack allied naval task groups or merchant convoys.

coming years. The *Los Angeles*-class nuclear-powered attack submarines, already under construction, are designed to seek out and destroy enemy submarines. The new Trident submarines are being developed to replace the older Polaris/Poseidon strategic missile submarines in the 1980s.

A small carrier or sea control ship is planned to carry advanced helicopters and fixed-wing aircraft. In time of war, these ships will escort merchant ships and amphibious task forces. A vertical/short take-off and landing fighter, the XFV-12, is under development by the Navy and industry for possible use aboard the sea control ship. However, the first sea control ships will use the AV-8 Harrier, now flown by the Marine Corps, as well as helicopters.

This is an artist's impression of the XVF-12 fighter designed to operate from small aircraft carriers or sea control ships. The jet-propelled aircraft will take off vertically or with only a short deck run and land in a similar manner. These VSTOL features will limit the flight performance of the aircraft, but its ability to operate from small ships and an "open spot" ashore will make it valuable in naval operations.

Another new ship series to be built in the future is the patrol frigate, an escort-type ship with antiaircraft and antiship missiles. This ship will help defend convoys against enemy patrol boats, aircraft, and submarines equipped with guided missiles.

Hydrofoils—craft which "fly" along the water on underwater wings called foils—will have wide use in the future Navy. The U.S. Navy has been experimenting with this concept for more than a decade, and it is now planning a series of guided missile patrol hydrofoils called PHMs. These 170-ton craft will be followed by a big 750-ton developmental hydrofoil (the DBH), a fast, 50-to-90-knot model (the DFH), and, possibly; by hydrofoil escort ships.

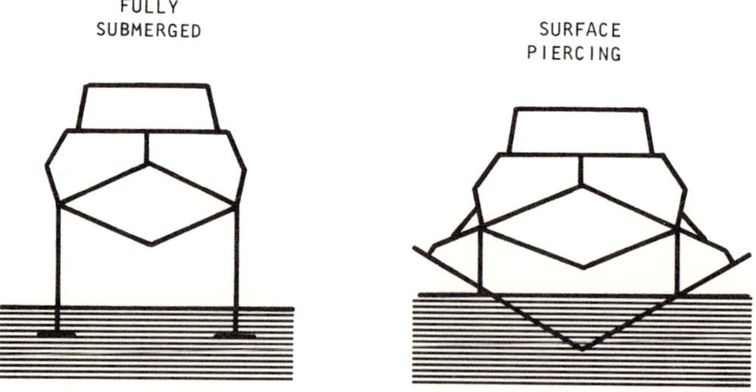

Small, fast-moving hydrofoil missile boats are being built by
the U.S. Navy for a variety of tasks.          *Boeing Company*



The Navy is also interested in air cushion vehicles (ACV)
and surface effect ships (SES) that ride *over* the surface of the
water on a cushion of air. The ACV has flexible "skirts"
surrounding the vehicle's base to trap the air cushion, while the

Surface effect ships, which ride on a "bubble" of air and skim over the surface of the water, will provide the Navy with high-speed ships with a transocean capability. Two missions for such ships that are under consideration provide for multi-mission escort ships (above), which will have advanced weapons and operate helicopters. The VSTOL aircraft, and amphibious assault/support ships (below), will be able to send Marines ashore by landing craft and helicopters, and also provide supplies to other ships.

SES has rigid sidewalls that penetrate the water surface to help hold the cushion or air bubble. These new craft are being built and tested for duty as patrol boats, landing craft, and, eventually, sub killers.

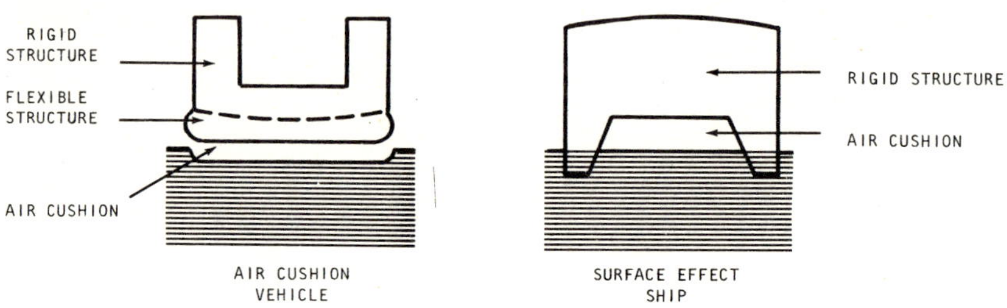


Another idea for the future is the small waterplane area twin hull (SWATH) ship, which will use two fully submerged, torpedo-shaped hulls to support the ship's armament and house its electronics and propulsion. The SWATH ship will be very stable, even in rough water, and it will be an ideal vehicle for antisubmarine warfare.

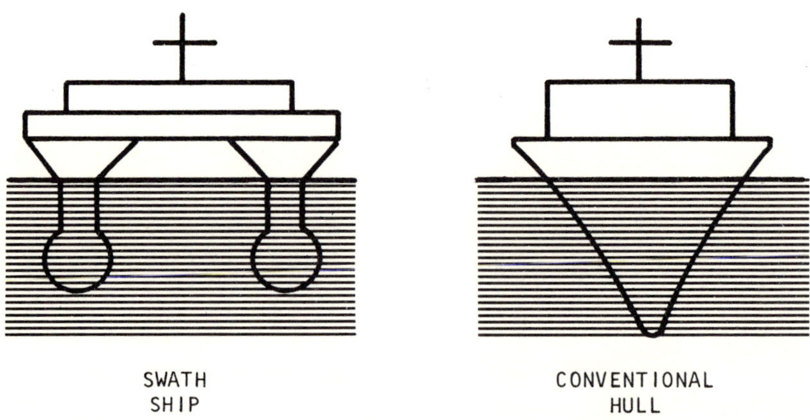

The small waterplane area twin hull (SWATH) ship is being considered for some Navy missions. The twin, cylinder-type hulls contain engines and other machinery. By placing all of the "hull" underwater, the ship's drag while pushing through the water is reduced and hence more speed can be developed for a given amount of power over a conventional ship.

The long era of the big gun and muscle is almost over. Naval operations are seeing the twilight of the all-gun ship; the Navy's new warships have at least some guided missiles aboard. The need for brute strength and sweat on deck and in the engine room is being eliminated by automation and new equipment meant to make life easier for ship crews. The human eyeball which strained to see telltale signs of enemy submarines or aircraft in the distance now scans radarscopes, electronic displays, and instrument panels. Today's Navy is a partnership between man and the transistor.

The new missions, ships, aircraft, and equipment will require the best the United States can produce in crews and operators. America's youth are receiving the best educations in history, and this knowledge and skill are needed in tomorrow's Navy. Even today, the fleet is receiving better trained and more highly qualified people than ever before. The older chief petty officers with World War II and Korean War experience are greeting "youngsters" into their ranks.

The younger Navymen and women are the technicians who must operate and maintain the sophisticated weapons, electronics, and other systems which are coming into the fleet. This doesn't mean the older, traditional specialties are no longer needed. Men must still signal, steer, handle lines, run boats and machinery, and perform the age-old tasks of the sea. But standing alongside the classic sailors are increasing numbers of bright, dedicated, young men and women who, while belonging to today's Navy, are also the first real members of the Navy of the twenty-first century.

# APPENDICES

# APPENDIX A  Navy Jobs

While the designations of many traditional Navy jobs imply that they are "for men only," many Navy positions are open to women as well. The prime exceptions are those directly involving combat duty which are, by law, restricted to men.

## Aerographer's Mate

Weather predictions are vital to the safe and efficient operation of the Navy's aircraft, surface ships, and land installations. The weatherman of the Navy is the aerographer's mate—observer, forecaster, and distributor of accurate weather information.

## Air Controlman

The safe, orderly, and speedy flow of air traffic is essential for the success of naval air operations. Air controlmen provide assistance to naval pilots in landing and taking off their planes.

## Aircrew Survival Equipmentman

Parachutes are the lifesaving equipment of aircraft crewmen when they have to bail out. In time of disaster, a parachute may be the only means of delivering badly needed medicines, food, and other supplies to isolated victims. The Navy's aircrew survival equipmentman has the highly responsible job of keeping parachutes and other aviation survival equipment in perfect working condition.

## Aviation Antisubmarine Warfare Operator

These operators perform as flight crew members to identify submarines by analyzing data obtained from antisubmarine sensor equipment. They perform flight associated maintenance for airborne electronic equipment. In addition, they supervise and train team members, using both ground training devices and in-flight training.

## Aviation Antisubmarine Warfare Technician

The Navy's antisubmarine warfare effort relies on naval aviation as well as surface and submarine forces for its knockout punch. The aviation antisubmarine warfare technician is an important member of this powerful team, keeping airborne electronic systems and equipment functioning during antisubmarine operations.

## Aviation Boatswain's Mate

Launching naval aircraft quickly and safely from ships or land requires deft handling on the part of the ground crew or deck force. Aviation boatswain's mates have a responsible part in these operations and in the handling of planes prior to takeoff and after landing.

## Aviation Electrician's Mate

The maze of electrical mechanisms and connecting wiring in modern aircraft requires expert care. Aviation electrician's mates keep this equipment in top operating condition.

## Aviation Electronics Technician

Modern naval aircraft operating from carriers, cruisers, or land bases depend upon radio, radar, and other electronic devices for rapid communications, efficient navigation, controlled landing approaches, detection of and guidance to enemy or other objectives, and reduction of effectiveness of enemy equipment and tactics. Aviation electronics technicians are responsible for the operating condition of this equipment.

## Aviation Fire Control Technician

The firing of guns on Navy combat planes is controlled by highly complex equipment. Aviation fire control technicians keep this equipment in operating condition through systematic maintenance and repair.

## Aviation Machinist's Mate

The aviation machinist's mates maintain reciprocal and jet engines. He will most likely be assigned to billets concerning the maintenance of turbojet aircraft engines and associated equipment. An aviation machinist's mate may be attached to any one of the several types of aircraft maintenance activities.

## Aviation Maintenance Administrationman

Aviation maintenance administrationmen perform administrative, management, and clerical duties. The vastness of the aviation branch of the Navy today requires specialists such as this in implementing and supporting the Aircraft Maintenance Program.

## Aviation Ordnanceman

Modern Navy aircraft have increased the range of naval weapons from a few miles to hundreds of miles. They carry guns, bombs, torpedoes, rockets, and missiles to attack the enemy on the sea, under the sea, in the air, and on the land. One of the specialists responsible for the perfect working order of armament on Navy planes is the aviation ordnanceman.

## Aviation Storekeeper

The large-scale development of naval aviation brought with it the attendant problem of supply. The various types of naval aircraft with their specialized parts, equipment, and supplies brought about a need for personnel especially trained in this field. The aviation storekeeper has met this need and performs the duties required for aviation supply.

## Aviation Structural Mechanic

Aircraft wings, fuselage, tail, control surfaces, landing gear, and attending mechanisms require skillful maintenance and repair. Aviation structural mechanics perform these jobs, working with a variety of metals, alloys, and plastics.

## Aviation Support Equipment Technician

The maintenance and repair of aviation support equipment is of vital importance at U.S. naval air stations and aboard aircraft carriers. Aviation support equipment is commonly referred to as the "yellow equipment." This equipment must be kept operational at all times, for without some of these equipments, the aircraft would not be able to fly. The aviation support equipment is attached to the plane only when the aircraft is on the ground or deck. The responsibility of the aviation support equipment technician is to keep the equipment working.

## Boatswain's Mate

The Navy develops master seamen—persons skilled in all phases of seamanship and in the handling of deck force personnel. They are the masters of many trades, able to perform almost any task connected with the operation of small boats, navigation, entering or leaving port, storing cargo, handling ropes and lines, and many others. These master seamen are the boatswain's mates.

## Boilermaker

When marine boilers and heat exchanges require major repair or overhaul, it is the boilermaker who is called on to do the work. The boilermaker is a highly trained repairman who effectively maintains the equipment which keep the Navy's ships moving.

## Boiler Technician

The propelling agent of our large naval ships is steam. Efficient operation, maintenance, and repair of marine boilers are essential for effective production of steam power. The Navy leans heavily upon its boiler technicians to keep its ships moving.

## Builder

Advanced base operations require the construction of many buildings, piers, waterfront structures, bridges, towers, and similar projects. Builders—part of the "seabee" organization—play an important part in the erection, maintenance, and repair of such structures.

## Commissaryman

Navy kitchens share largely in the responsibility for maintaining at high level the health and morale of Navy personnel. These kitchens are operated by commissarymen responsible for menus and the care of food supplies; for baking, broiling, and frying; for mixing, seasoning, and flavoring; and other culinary preparations contributing toward wholesome, satisfying meals.

## Communications Technician

The secrets of a nation are only as secure as its communication system. Communications technicians perform highly specialized duties concerned with special communications, special operations, and communications security.

## Construction Electrician

Advanced bases require the construction of roads, barracks, airfields, hospitals, shops, and warehouses. Construction electricians are responsible for all power production and electrical work essential to the establishment and operation of these bases.

## Construction Mechanic

Maintaining automotive and heavy construction equipment in efficient operating condition requires the skills of highly trained technicians. Construction mechanics perform these essential tasks on both diesel and gasoline internal-combustion engines.

## Data Processing Technician

Like any large commercial enterprise, the U.S. Navy has an extensive accounting system. Complete records must be maintained for every person with reference to health, pay, and qualifications; for every ship and station with reference to allotment of personnel, receipt and transfer of supplies and personnel, and disbursement of money; and for every piece of equipment the Navy owns. To keep these records up to date, to ensure their accuracy, and to make the tabulated information immediately available in any form which may be desired, the Navy makes wide use of a wide range of data processing equipment. Data processing technicians operate and maintain this equipment.

## Data Systems Technician

Naval warfare in this age of atomic power and ballistic missiles requires quick and correct answers to complicated mathematical problems. These answers are obtained by use of electronic digital computers which accomplish in minutes what a man might take days to do. The data systems technician keeps this vital equipment in operation.

## Dental Technician

In the Navy, as in civil life, health is one of the most important factors in efficient job performance. Proper care of the teeth is essential to the total health picture. The dental technician assists the dental officer in all phases of dental work.

## Disbursing Clerk

The construction, maintenance, and operation of great ocean and air fleets have put the Navy into business on a scale comparable with the largest civilian industrial enterprises. The Navy's payroll is one of the largest in the world, and the Navy is a proportionately large consumer of the world's goods. Regular servicing of this payroll, financial transactions involved in procuring materials and services, and the related accounting functions are the job of Navy disbursing clerks.

## Electrician's Mate

The operation and repair of the ship's electrical power plant and electrical equipment is the responsibility of the electrician's mate.

## Electronics Technician

All of the electronics equipment used in the Navy to send and receive messages, detect enemy planes and ships, and determine distance of targets requires continuous checking and repairing. Electronics technicians are highly trained personnel who take care of this equipment.

## Electronics Warfare Technician

Electronics warfare technicians repair and maintain the Navy's electronic warfare equipment.

## Engineering Aid

"This will be the airstrip, the road will be cut through those trees and down there we will put the administration building." From this beginning the Navy's engineering aids start their work so the engineers will have data for developing construction plans.

## Engineman

The diesel and gasoline engine have a tremendously important role in powering the ships and small craft of the Navy. These engines must be properly maintained, repaired, and operated. The Navy engineman's work centers around these important jobs.

## Equipment Operator

The Navy uses large, self-powered equipment in its work of construction, repair, salvage, and excavation. The bulldozer is a well-known example of this type of equipment. The equipment operator operates the bulldozer and other types of construction equipment.

## Fire Control Technician

Extremely complicated electronic, electrical, hydraulic, and mechanical equipment is required to  compute and resolve the many factors which influence the accuracy of naval guided missiles, gunfire, and underwater weapons. Maintenance and repair of the equipment is the prime responsibility of highly skilled specialists—the fire control technicians.

## Gas Turbine Systems Technician

The new gas turbine systems technician rating will give the Navy skilled personnel to run the sophisticated propulsion plants of the future. In addition to his responsibility for heavy machinery operation, the gas turbine systems technician will maintain solid state logic circuitry used to control the entire main propulsion system.

## Gunner's Mate

Navy ships equipped with various guns have long been protectors against enemy aggressors. Navy's gunner's mates operate, maintain, and repair all gunnery equipment as well as handle ammunition used on Navy ships. Gunner's mates also maintain guided missiles on surface ships.

## Hospital Corpsman

Much of the credit for the good health of Navy personnel is due to the work of the hospital corpsmen. They are the Navy's pharmacists, medical technicians, and first-aid men.

## Hull Maintenance Technician

Hull maintenance technicians perform the jobs previously filled by damage controlmen and shipfitters. They are responsible for safety and survival devices aboard ship. In addition, they maintain and repair ship hulls, fittings, piping systems, and machinery.

## Illustrator Draftsman

Accurate, clear mechanical drawings, blueprints, charts, and illustrations are essential in the planning and completion of construction projects and for other purposes in the Navy. Skilled illustrator draftsmen develop and produce these highly important aids.

## Instrumentman

The Navy uses large numbers of meters and gauges, watches and clocks, typewriters, adding machines, and other types of office machines. To maintain these many and varied machines in good working order requires the services of instrumentmen who are highly skilled in a number of specialized activities.

## Interior Communications Technician

Communications systems throughout a Navy ship make a vital contribution to her operating efficiency. The operation and repair of electronics devices used in the ship's interior communications systems, public-address systems, electric megaphones, and other announcing equipment are the responsibility of the I.C. electricians.

## Journalist

The journalist plays an important part in maintaining high Navy morale through the dissemination of news and in keeping the general public informed as to the developments, accomplishments, and policies of the Navy. This is done through ship and station newspapers, bulletins, pamphlets, news releases, and radio scripts.

## Legalman

Legalmen are trained in court reporting, claims matters, investigations, legal administration, and legal research.

## Lithographer

The Navy's presses reproduce thousands of printed items used in the Navy's many functions: recruiting posters and career counseling aids;

manuals, training materials, and record forms used in administering and instructing naval personnel; and bulletins, magazines, and newspapers to inform and entertain.

## Machinery Repairman

The replacement of parts and the repair of machinery on shipboard and ashore is done in the Navy's machine shops. Machinery repairmen do this work and operate the shops.

## Machinist's Mate

Continuous operation of the many engines, compressors, gears, refrigerating, air-conditioning, gas-operating equipment, and other types of machinery aboard modern Navy vessels and at various shore stations depends upon the skill of specially trained technicians. Machinist's mates are the technicians responsible for the operation, maintenance, and repair of this machinery.

## Master-at-Arms

The master-at-arms rating is for law enforcement specialists, and involves shore patrol, Armed Forces police work, security, and criminal investigation.

## Mineman

Sea mines are silent, unseen sentries maintaining defensive sea blockades which once required whole fleets of surface ships. Mines are offensive weapons of sea warfare, with highly complicated firing mechanisms. Assembling and repairing these weapons is the job of the mineman.

## Missile Technician

Guided missiles appear on increasing numbers of Navy warships. Missile technicians have the important job of maintaining and repairing these weapons.

## Molder

Many of the metal parts used in the repair of ships, guns, and other equipment are machined in Navy shops from rough castings. It is the job of molders to make these rough castings as they are needed.

## Musician

As members of Navy bands and orchestras, musicians provide entertainment in every corner of the world. The U.S. Navy Band at Washington and the U.S. Naval Academy Band at Annapolis are examples of Navy musical organizations.

## Navy Counselor

Navy counselors work in the fields of recruiting and career counseling. This job has gained new significance with the end of the draft and adoption of an all-volunteer military force.

## Ocean Systems Technician

The Navy requires a constant supply of oceanographic data for purposes of research and development and the support of the forces afloat and other operational commands. The collection and interpretation of these data require the special skills of the ocean systems technician.

## Operations Specialist

Radar—an electronic device to determine the presence and location of an object—is used extensively in navigation and maneuvering, in recognition and identification, in searching for and following the movements of other ships and aircraft. The responsibility of the operations specialist is to operate and to interpret the information received from radar.

## Opticalman

Modern marine navigation and aviation owe much of their efficiency to the use of scientifically accurate optical instruments such as octants and sextants for navigation, range finders and sights for gunnery, and binoculars and telescopes for magnification. Keeping these instruments in good working order is the job of the opticalman.

## Patternmaker

Foundries at naval shipyards and on repair ships are continually called upon to produce castings of a specialized nature. The patternmaker is the important link between the draftsmen who make the drawings and the foundrymen who will produce the castings. The patternmaker creates exact patterns to be used by the foundrymen to form molds for making the castings.

## Personnelman

Personnelmen perform personnel administration duties such as counseling enlisted personnel about training, promotion requirements, educational opportunities, and the benefits and advantages of a Navy career; conducting tests and interviews in various personnel programs, maintaining publications and directives regarding enlisted personnel, and performing clerical duties related to personnel administration.

## Photographer's Mate

Naval activities in peace and war are carefully recorded visually by motion pictures and still camera photographs. These pictorial records of historical and newsworthy events aboard ship and at shore stations are made by photographer's mates.

## Photographic Intelligence Technician

There is an important need for continued availability of precise and extremely detailed intelligence information covering all aspects of target areas in nations which may become enemies of the United States. The collection and presentation of this information requires the special skills of the photographic intelligence technician.

## Postal Clerk

An efficient Navy postal service is vital to the smooth functioning of this global arm of the national defense forces. Not only must official mail be delivered promptly to the commanding officers concerned, but the personal mail of Navymen—the link with their loved ones—must be equally dependable, or morale suffers. The Navy postal organization provides every service that civilian post offices provide.

## Quartermaster

The safety of ships at sea depends to a great extent on skillful navigation; the vigilance with which lookouts for enemy ships and aircraft, water traffic, and natural obstacles is maintained; and the proficiency with which signals are exchanged with other ships and the shore. The quartermaster performs or assists in the performance of these duties.

## Radioman

All naval actions require complex teamwork, sometimes involving hundreds of individual units. One of the major factors in the success of these operations is accurate and speedy transmission of radio messages. Radiomen transmit and receive these messages and keep their equipment in good operating condition.

## Ship's Serviceman

Wherever Navy personnel may be, afloat or ashore, they can obtain the services and commodities available in civilian life—from having their hair cut to getting their shoes repaired and from purchasing soap and razor blades to buying ice cream and gifts. These services are provided by the ship's servicemen.

## Signalman

The maneuvers at sea depend upon rapid and accurate communications. The safety of ships depends to a great extent on the vigilance of lookouts for enemy ships and aircraft, sea traffic, and natural obstacles. These duties are performed by signalmen who send and receive messages by flashing light, semaphore, and flag hoist, and serve as lookouts.

## Sonar Technician

The Navy uses sonar—underwater listening devices—to determine what is under the water as well as what is on the surface, detecting reefs in uncharted waters, and discovering the presence of enemy submarines, surface ships, or other submerged objects. The operation and care of sonar equipment is the responsibility of the sonar technician.

## Steelworker

The pieces of structural steel that form the frames of naval hangars, radio towers, storage tanks, pontoons, dry docks, bridges, and other structures must be hoisted into place, bolted together temporarily, trued, and then joined together solidly. This is the primary job of the steelworker.

## Steward

Stewards purchase, prepare, and serve food for the wardrooms (officers' messes). They are skilled cooks and bakers and the custodians of officers' quarters.

## Storekeeper

Navy ships and shore bases require a constant supply of clothing, spare parts, technical items, and other essential supplies. Providing and accounting for these materials are the main responsibilities of the storekeeper.

## Torpedoman's Mate

Torpedoes are highly sophisticated, electronic and mechanical, self-propelled, passive or target-seeking underwater weapons. Their effectiveness depends upon the proper preparation, maintenance, loading, and firing by torpedoman's mates.

## Tradevman

The training of Navy personnel requires highly specialized equipment. Various types of training aids and training devices are used to simulate actual operating conditions. The success of this phase of the Navy training program depends upon how well the tradevmen teach others to use it. The term "tradev" is derived from training devices.

## Utilitiesman

Water, light, heat, power generating, and sewage disposal equipment must be provided for advanced bases and large continental shore bases. The utilitiesmen install, operate, maintain, and repair these important facilities.

## Yeoman

Communications among activities within the Navy, with other government agencies, private industry, and individuals is necessary in conducting naval affairs. Vast numbers of letters, messages, and records must be prepared in procuring and utilizing the personnel and material required to operate the fleets. Navy yeomen perform these office duties.

# Navy Aircraft

## FIXED-WING, CARRIER-BASED

| | | |
|---|---|---|
| **A-7** | **Corsair II** | Standard Navy carrier-based light attack aircraft. Engines—1 jet; crew—1. |
| **F-8** | **Crusader** | Fighter used by Navy aboard smaller aircraft carriers; being phased out as older carriers are retired: RF-8 Crusader is reconnaissance version. Engines—1 jet; crew—1. |
| **C-2** | **Greyhound** | Improved carrier on-board delivery (COD) aircraft. Engines—2 turboprop; crew—3. |
| **AV-8** | **Harrier** | Vertical/short take-off and landing (VSTOL) aircraft used by Marines for support of ground troops; a few are being assigned to small carriers called sea control ships. Engines—2 jet; crew—1. |
| **E-2** | **Hawkeye** | Navy electronic warfare aircraft. Engines—2 turboprop; crew—5. |
| **A-6** | **Intruder** | All-weather, day-night attack plane capable of carrying several tons of bombs or missiles; used |

|  |  | by Navy and Marines; KA-6 Intruder is an aerial tanker used by Navy to refuel other carrier aircraft. Engines—2 jet; crew—2. |
|---|---|---|
| **F-4** | **Phantom II** | Standard Navy and Marine fighter; can carry bombs or missiles for ground and ship attack; considered one of best fighters in non-Communist world. Marines also fly RF-4 reconnaissance version. Engines—2 jet; crew—2. |
| **EA-6** | **Prowler** | Electronics countermeasure aircraft used to detect and jam enemy radars; used by Navy and Marines. Engines—2 jet; crew—2 or 4. |
| **A-4** | **Skyhawk** | Light attack aircraft flown primarily by Marines. Two-seat TA-4 Skyhawk is used for pilot training. Engines—1 jet; crew—1. |
| **F-14** | **Tomcat** | Advanced carrier-based fighter for Navy; joined the fleet in 1973; has variable-sweep wings that automatically adjust as plane maneuvers; "hottest" fighter in non-Communist world. Engines—2 jet; crew—2. |
| **E-1** | **Tracer** | Used in electronic warfare missions. Engines—2 piston; crew—4. |
| **S-2** | **Tracker** | Navy carrier-based antisubmarine aircraft; carries radar, droppable sonar buoys, magnetic detectors, and other equipment for hunting subs, as well as torpedoes and rockets for attacking them. To be replaced by S-3 Viking. Engines—2 piston; crew—4. |
| **C-1** | **Trader** | Cargo version of S-2 Tracker; used to deliver passengers and cargo to ships at sea. Engines—2 piston; crew—2 plus 9 others. |
| **RA-5** | **Vigilante** | Large carrier-based reconnaissance plane for gathering photographic, infrared, and electronic intelligence. Engines—2 jet; crew—2. |
| **S-3** | **Viking** | Advanced Navy antisubmarine aircraft for use from carriers. Engines—2 jet; crew—4. |

# FIXED-WING, LAND-BASED

| | | |
|---|---|---|
| **TC-4** | **Academe** | Navigation training aircraft. Engines—2 turboprop; crew—1 plus 6 students. |
| **0-1** | **Bird Dog** | Small observation aircraft. Engines—1 piston; crew—2. |
| **0V-10** | **Bronco** | Multipurpose armed reconnaissance aircraft. Engines—2 turboprop; crew—2. |
| **T-2** | **Buckeye** | Basic trainer used for carrier landings. Engines—1 jet; crew—2. |
| **T-29** | **Flying Classroom** | Navigation training aircraft. Engines—2 piston; crew—4 plus 10 students. |
| **C-130** | **Hercules** | Transport aircraft used for carrying cargo or troops. Marines use KC-130 as aerial tanker. Engines—4 turboprop; crew—7. |
| **C-118** | **Liftmaster** | Similar to DC-6; used for troops, passengers, or medical litters. Engines—4 piston; crew—6 plus 76 troops. |
| **P-2** | **Neptune** | Patrol aircraft flown mainly by Naval Reserve. Engines—2 piston; crew—9. |
| **C-9** | **Nightingale** | Medium transport similar to DC-9. Engines—2 jet; crew—2 plus 90 troops. |
| **P-3** | **Orion** | Long-range reconnaissance and antisubmarine patrol aircraft; flown by Navy from land bases in the United States and overseas. Fitted with sophisticated submarine detection equipment and weapons. Engines—4 turboprop; crew—10. |
| **T-39** | **Sabreliner** | Radar training aircraft; CT-39 is transport version. Engines—2 jet; crew—2 to 5. |
| **C-131** | **Samaritan** | Personnel and cargo transport aircraft similar to Convair 340 or 440. Engines—2 piston; crew—3 plus 44. |
| **T-1** | **Sea Star** | Jet training aircraft. Engines—1 jet; crew—2. |

| C-117 | **Skytrain** | Similar to classic DC-3; C-47 older version; used for passenger transport. Engines—2 piston; crew—3 plus 8 to 35 passengers. |

# HELICOPTERS

| UH-1 | **Iroquois** | Generally known as "Huey" (for old HU-IE designation) or "Sea Wolf"; used as gunship and for observation, primarily by Marines. Crew—2 plus 7 passengers. |
| AH-1 | **Seacobra** | Attack gunship used for supporting Marines on the ground. Crew—2. |
| UH-34 | **Seahorse** | Utility helicopter being phased out of service. Crew—2 to 4 plus 14 passengers. |
| SH-3 | **Sea King** | Large, two-engine antisubmarine helicopter used aboard aircraft carriers for antisubmarine missions; HH-3 version is used in rescue operations. Crew—4. |
| H-46 | **Sea Knight** | CH-46 is flown by Marines to carry troops and cargo; UH-46 operated by Navy to transfer cargo between ships at sea in vertical replenishment operations. Crew—3 plus 25 to 33 troops. |
| SH-2 | **Sealite** | Small antisubmarine helicopter used aboard destroyer-type ships. Crew—4. |
| TH-57 | **Searanger** | Training helicopter. Crew—5. |
| UH-2 | **Seasprite** | Small utility helicopter used for transferring passengers between ships; HH-2 version engages in rescue operations. Crew—2 to 3 plus 10 passengers. |
| CH-53 | **Sea Stallion** | Large Marine cargo helicopter for carrying up to 38 troops; RH-53 version used by the Navy for sweeping mines. Crew—3 to 6. |

# APPENDIX C
# Navy Ship Types

## COMBATANT SHIPS

### Warships
#### Aircraft Carriers:

| | |
|---|---|
| CV | Aircraft Carrier |
| CVA | Attack Aircraft Carrier |
| CVAN | Attack Aircraft Carrier (nuclear propulsion) |
| CVS | ASW Aircraft Carrier |

#### Surface Combatants:

| | |
|---|---|
| BB | Battleship |
| CA | Heavy Cruiser |
| CG | Guided Missile Cruiser |
| CGN | Guided Missile Cruiser (nuclear propulsion) |
| CL | Light Cruiser |
| CLG | Guided Missile Light Cruiser |
| DD | Destroyer |
| DDG | Guided Missile Destroyer |
| DDR | Radar Picket Destroyer |
| DL | Frigate |
| DLG | Guided Missile Frigate |
| DLGN | Guided Missile Frigate (nuclear propulsion) |

#### Ocean Escorts:

| | |
|---|---|
| DE | Escort Ship |
| DEG | Guided Missile Escort Ship |
| DER | Radar Picket Escort Ship |

#### Submarines:

| | |
|---|---|
| SS | Submarine |

| | |
|---|---|
| SSBN | Fleet Ballistic Missile Submarine (nuclear propulsion) |
| SSG | Guided Missile Submarine |
| SSN | Submarine (nuclear propulsion) |

**Patrol Ships:**

| | |
|---|---|
| PCE | Patrol Escort |
| PCER | Patrol Rescue Escort |
| PG | Patrol Gunboat |

## Amphibious Warfare Ships

| | |
|---|---|
| LCC | Amphibious Command Ship |
| LFR | Inshore Fire Support Ship |
| LFS | Amphibious Fire Support Ship |
| LHA | Amphibious Assault Ship (general purpose) |
| LKA | Amphibious Cargo Ship |
| LPA | Amphibious Transport |
| LPD | Amphibious Transport Dock |
| LPH | Amphibious Assault Ship |
| LPR | Amphibious Transport (small) |
| LPSS | Amphibious Transport Submarine |
| LSD | Dock Landing Ship |
| LST | Tank Landing Ship |

## Mine Warfare Ships

| | |
|---|---|
| MCS | Mine Countermeasures Ship |
| MSC | Minesweeper, Coastal (nonmagnetic) |
| MSF | Minesweeper, Fleet (steel hulled) |
| MSO | Minesweeper, Ocean (nonmagnetic) |

# COMBATANT CRAFT

## Patrol Craft

| | |
|---|---|
| PCH | Patrol Craft (hydrofoil) |
| PGH | Patrol Gunboat (hydrofoil) |
| PTF | Fast Patrol Craft |

## Landing Craft

| | |
|---|---|
| LCA | Landing Craft, Assault |
| LCM | Landing Craft, Mechanized |
| LCPL | Landing Craft, Personnel, Large |
| LCPR | Landing Craft, Personnel, Ramped |
| LCU | Landing Craft, Utility |
| LCVP | Landing Craft, Vehicle, Personnel |
| LWT | Amphibious Warping Tug |

## Mine Countermeasures Craft

| | |
|---|---|
| MSB | Minesweeping Boat |
| MSD | Minesweeper, Drone |
| MSI | Minesweeper, Inshore |
| MSL | Minesweeping Launch |
| MSM | Minesweeper, River |
| MSR | Minesweeper, Patrol |
| MSS | Minesweeper, Special |

## Riverine Warfare Craft

| | |
|---|---|
| ASPB | Assault Support Patrol Boat |
| ATC | Armored Troop Carrier |
| CCB | Command and Control Boat |

| | | | |
|---|---|---|---|
| MON | Monitor | LSSC | Light SEAL Support Craft |
| PBR | River Patrol Boat | MSSC | Medium SEAL Support Craft |
| PCF | Patrol Craft, Inshore | SDV | Swimmer Delivery Vehicle |
| QFB | Quiet Fast Boat | | |
| RUC | Riverine Utility Craft | | |
| STAB | Strike Assault Boat | | |

**SEAL Support Craft**

LCSR  Landing Craft Swimmer Reconnaissance

**Mobile Inshore Underseas Warfare (MIUW) Craft**

MAC  MIUW Attack Craft

# AUXILIARY SHIPS

| | | | |
|---|---|---|---|
| AD | Destroyer Tender | AK | Cargo Ship |
| ADG | Degaussing Ship | AKD | Cargo Ship Dock |
| AE | Ammunition Ship | AKL | Light Cargo Ship |
| AF | Store Ship | AKR | Vehicle Cargo Ship |
| AFS | Combat Store Ship | AKS | Stores Issue Ship |
| AG | Miscellaneous | AKV | Cargo Ship and Aircraft Ferry |
| AGDE | Escort Research Ship | ANL | Net Laying Ship |
| AGEH | Hydrofoil Research Ship | AO | Oiler |
| AGER | Environmental Research Ship | AOE | Fast Combat Support Ship |
| AGF | Miscevwous Command Ship | AOG | Gasoline Tanker |
| AGM | Missile Range Instrumentation Ship | AOR | Replenishment Oiler |
| | | AP | Transport |
| AGMR | Major Communications Relay Ship | APB | Self-propelled Barracks Ship |
| AGOR | Oceanographic Research Ship | AR | Repair Ship |
| | | ARB | Battle Damage Repair Ship |
| AGP | Patrol Craft Tender | ARC | Cable Repairing Ship |
| AGR | Radar Picket Ship | ARG | Internal Combustion Engine Repair Ship |
| AGS | Surveying Ship | | |
| AGSS | Auxiliary Submarine | ARL | Landing Craft Repair Ship |
| AGTR | Technical Research Ship | | |
| | | ARS | Salvage Ship |
| AH | Hospital Ship | ARSD | Salvage Lifting Ship |

| | | | | |
|---|---|---|---|---|
| ARST | Salvage Craft Tender | | ATSS | Auxiliary Training Submarine |
| ARVA | Aircraft Repair Ship (Aircraft) | | AV | Seaplane Tender |
| ARVE | Aircraft Repair Ship (Engine) | | AVM | Guided Missile Ship |
| ARVH | Aircraft Repair Ship (Helicopter) | | AVS | Aviation Supply Ship |
| AS | Submarine Tender | | AVT | Auxiliary Aircraft Transport |
| ASR | Submarine Rescue Ship | | AW | Distilling Ship |
| ATA | Auxiliary Ocean Tug | | CVT | Training Aircraft Carrier |
| ATF | Fleet Ocean Tug | | FDL | Fast Deployment Logistic Ship |
| ATS | Salvage Tug | | | |

# SERVICE CRAFT

| | | | | |
|---|---|---|---|---|
| AFDB | Large Auxiliary Floating Dry Dock | | YAG | Miscellaneous Auxiliary (self-propelled) |
| AFDL | Small Auxiliary Floating Dry Dock | | YC | Open Lighter |
| AFDM | Medium Auxiliary Floating Dry Dock | | YCF | Car Float |
| APL | Barracks Craft | | YCV | Aircraft Transportation Lighter |
| ARD | Auxiliary Repair Dry Dock | | YD | Floating Crane |
| ARDM | Medium Auxiliary Repair Dry Dock | | YDT | Diving Tender |
| DSRV | Deep Submergence Rescue Vehicle | | YF | Covered Lighter (self-propelled) |
| DSV | Deep Submergence Vehicle | | YFB | Ferryboat or Launch (self-propelled) |
| IX | Unclassified Miscellaneous | | YFD | Yard Floating Dry Dock |
| NR | Submersible Research Vehicle (nuclear propulsion) | | YFN | Covered Lighter |
| | | | YFNB | Large Covered Lighter |
| SST | Target and Training Submarine (self-propelled) | | YFND | Dry Dock Companion Craft |
| | | | YFNX | Lighter (special purpose) |
| X | Submersible Craft (self-propelled) | | YFP | Floating Power Barge |
| | | | YFR | Refrigerated Covered Lighter (self-propelled) |
| | | | YFRN | Refrigerated Covered Lighter |

| | |
|---|---|
| YFRT | Covered Lighter (Range Tender) (self-propelled) |
| YFU | Harbor Utility Craft (self-propelled) |
| YG | Garbage Lighter (self-propelled) |
| YGN | Garbage Lighter (non-self-propelled) |
| YHLC | Salvage Lift Craft, Heavy |
| YLLC | Salvage Lift Craft, Light (self-propelled) |
| YM | Dredge (self-propelled) |
| YMLC | Salvage Lift Craft, Medium |
| YNG | Gate Craft |
| YO | Fuel Oil Barge (self-propelled) |
| YOG | Gasoline Barge (self-propelled) |
| YOGN | Gasoline Barge (non-self-propelled) |
| YON | Fuel Oil Barge (non-self-propelled) |
| YOS | Oil Storage Barge |
| YP | Patrol Craft (self-propelled) |
| YPD | Floating Pile Driver |
| YR | Floating Workshop |
| YRB | Repair and Berthing Barge |
| YRBM | Repair, Berthing, and Messing Barge |
| YRDH | Floating Dry Dock Workshop (Hull) |
| YRDM | Floating Dry Dock Workshop (Machine) |
| YRR | Radiological Repair Barge |
| YRST | Salvage Craft Tender |
| YSD | Seaplane Wrecking Derrick (self-propelled) |
| YSR | Sludge Removal Barge |
| YTB | Large Harbor Tug (self-propelled) |
| YTL | Small Harbor Tug (self-propelled) |
| YTM | Medium Harbor Tug (self-propelled) |
| YV | Drone Aircraft Catapult Control Craft |
| YW | Water Barge (self-propelled) |
| YWN | Water Barge (non-self-propelled) |